MACHINES IN MOTION

*Michael Faraday giving a
Christmas Lecture in the main
lecture theatre of the Royal
Institution in December, 1855.
The Prince Consort and the
young Princes Albert and Edward
are in the centre of the front row.
Courtesy of the Royal Institution.*

MACHINES IN MOTION

Leonard Maunder OBE, PhD, ScD, FIMechE, FEng

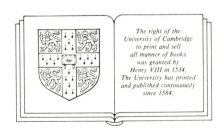

The right of the University of Cambridge to print and sell all manner of books was granted by Henry VIII in 1534. The University has printed and published continuously since 1584.

CAMBRIDGE UNIVERSITY PRESS

Cambridge

London   New York   New Rochelle

Melbourne   Sydney

Published by the Press Syndicate of the University of Cambridge
The Pitt Building, Trumpington Street, Cambridge CB2 1RP
32 East 57th Street, New York, NY 10022, USA
10 Stamford Road, Oakleigh, Melbourne 3166, Australia

First published 1986

Printed in Great Britain by Ebenezer Baylis, Worcester

British Library cataloguing in publication data

Maunder, Leonard
   Machines in motion
      1. Machinery   2. Mechanics
      I. Title
      621.8′11   TJ145

Library of Congress cataloguing in publication data

Maunder, Leonard.
   Machines in motion.

   Bibliography
   Includes index.
      1. Machinery.   2. Motion.   I. Title.
TJ145.M37 1986   621.8′11   86−9572

ISBN 0 521 30034 7

DS

To Moira

Contents

'. . . things in motion sooner catch the eye
Than what not stirs . . .'
ULYSSES' ADVICE TO ACHILLES, WILLIAM SHAKESPEARE.

Preface

An author who aims to translate the Christmas Lectures of the Royal Institution into a book is likely to discover that he has more to say in a preface than a decent brevity allows. He has at his disposal the illustrious history of the lectures, founded in 1826 and still going strong, the unique context of the Royal Institution, and a persistent temptation to dwell on the more intransigent aspects of the transfer from one medium to another. In the book I have tried to capture the spirit of the lectures by including photographs of actual events in the lecture theatre and during rehearsals: this material, together with printed slides, determined the main structure of the book, but at various points it has been complemented by fuller accounts than the spoken word could provide. A few references applying to each chapter are included for further reading.

Altogether about 150 experiments and demonstrations, 50 slides and 20 short films were presented in the six allotted hours, a programme that might still be going on were it not for the kindly but firm discipline of the BBC producer and his team. The subject of motion is of such wide interest and application that personal selection was inevitable, and it seemed best to follow a theme linking great discoveries and mechanical practice. It should not be supposed, of course, that the traffic of ideas goes in one direction only, for the struggle to develop new kinds of machines has contributed hugely to new discoveries. If the book serves to mark the strength of the flow in *both* directions, it will have served a useful purpose. But the main purpose is to interest the reader, particularly the younger reader, in a subject that will continue to attract investigation, invention and design into the foreseeable future.

ACKNOWLEDGEMENTS

I acknowledge with thanks the contributions of many people to the writing of the book. Mr Bill Coates and his staff at the Royal Institution, who supported the lectures with infinite resource, also repeated experiments for the written record. Mr Brian Johnson and his BBC team controlled the filming and suggested valuable improvements to the content, and Mr Martin Broadfoot kindly made available photographs he took during the live presentations. University colleagues showed great patience in response to endless enquiries, and the assistance of Mr A. Jameson during the lectures and Mr W. Burgess (who between them constructed many of the experimental models), Mr A. McGilligan (who took the laboratory photographs), Mr R. Adamson (who drew the illustrations) and Mrs S. Stone (who typed the manuscript) was invaluable. Many slides and models used in the lectures were loaned by other individuals and organizations, and every effort has been made to acknowledge the sources in the text.

Finally, my wife's partnership in the lectures and her patience during the writing of the book were indispensable.

1

DRIVING FORCES

Introduction

When the first Christmas Lectures were given at the Royal Institution in 1826, the audience assembled in the same historic and precipitious lecture theatre that is used today. A painting of one of Michael Faraday's lectures of that era, reproduced in the frontispiece, gives a good impression both of the arrangement and of the spirit of the times. Since no one could be absolutely sure what would happen during the experiments, a general air of anticipation, not to say adventure, must have been shared by audience and lecturer alike.

That much is still the same. But other things are different, both inside and outside the theatre. In 1826 there were no electric lights to switch on and off, the heating depended on a coal fire in a fireplace now vanished behind a screen, and it was advisable to be well wrapped up for a cold walk or horse-driven ride home. It was impossible to telephone friends, listen to the radio or watch television. Aeroplanes were out of the question, trains had just begun and no one owned a car or even a bicycle. Expeditions of more than a few miles were not to be undertaken lightly, as an earlier advertisement for the Manchester Flying Coach and its horses confirms: 'However incredible it may appear', went the proud announcement, 'this coach will actually (barring accidents) arrive in London four and a half days after leaving Manchester.'

Things have moved on. When we look back, the changes are astonishing. But when we look forward, it is difficult to see very far. If we make an inspired guess, and, if after some years we find we did not get it exactly right, we shall be in good company. Many well-informed observers in the past confidently ruled out railways, the motor-car, the aeroplane, space-flight . . . and so on. Developments that are familiar now were once but a gleam in the eye of engineers like Hero of Alexandria, James Watt, George Stephenson and Frank Whittle.

Machines, of course, are not the only things that move; everything else moves, too, from stars in space to electrons in atoms. A man sitting and dreaming on an equatorial beach is travelling about 700 miles per hour relative to an explorer camped at the North Pole. The atmosphere of the Earth may move gently over the land below or it may demolish buildings at speeds of over 100 miles per hour. The sudden shock of earthquakes can ravage whole regions, and the wind can whip up giant waves at sea. In the animal world, the ways in which birds fly, reptiles crawl and fish swim are not only amazingly varied but also contain valuable lessons for engineering applications. The study of human motion helps doctors and engineers to understand more about how we function, and it leads to better means of coping with disabilities. And behind the concepts of motion lie

a

Fig. 1.1(*a*) *Sir Charles Parsons, who invented the world's first practical steam turbine in 1884.*

Fig. 1.1(*b*) *The machine in question, now in the Science Museum, South Kensington. It developed 7.5 kW at 30 000 rev/min, and its overall length is just under 2 m.*

fascinating questions about how motion is controlled, be it in living or engineering machines. So, all in all, the subject of motion is a pretty wide one.

Let us start with a remarkable engineering success. In 1884, Charles Parsons, Fig. 1.1(*a*), was granted a famous patent for the world's first practical steam turbine. His original machine developed 7.5 kilowatts at the astonishing speed of 30 000 revolutions per minute, Fig. 1.1(*b*). Nowadays, giant turbines like that in Fig. 1.1(*c*) drive electrical generators that produce hundreds of megawatts from a single shaft, enough electricity for a large town; it is a remarkable fact that the entire electrical demand of England and Wales is supplied by about 350 behemoths of this size.

Parsons also applied steam turbines to ships. His team at Wallsend-on-Tyne built an historic little ship of 44 tons displacement, the

b

c

Fig. 1.1(*c*) *Construction of the low-pressure section of a modern steam turbine. The complete machine develops 660 MW at 3000 rev/min. Courtesy of NEI Parsons.*

Fig. 1.2 *The Turbinia, the world's first steam turbine driven ship, at speed. Now preserved at Newcastle upon Tyne, it reached 34 knots in 1897. Courtesy of NEI Parsons.*

Turbinia, which reached the unprecedented speed of 34 knots, Fig. 1.2. The Navy, however, was not impressed. To help their unbelief, Parsons on Turbinia infiltrated a great Naval Review at Spithead, where a vast fleet of warships was celebrating Queen Victoria's Diamond Jubilee. Records of the event differ, but rumour has it that he gate-crashed that mighty naval force. What is certain is that the effect was sensational. With Sir Charles in charge of the machinery, Turbinia raced down the lines of the assembled warships, leaving all behind. A few years later, steam turbines were adopted universally in ships of all the world's navies.

An ability to make good use of nature's materials and forces is the hallmark of the engineer. From the day some 6000 years ago, when the first wheel-and-axle rolled over a muddy track, to the present, when we depend on machines to maintain our civilizations, the imagination and skills of engineers have achieved spectacular and exciting advances. In a sense, we are still at a beginning, for human welfare throughout the world will never reach tolerable levels without massive new engineering contributions.

Two threads of history: theory and practice

If we include the invention of tools, the history of machines is almost as long as the history of man. By the time of the late Palaeolithic Age, a wide range of tools were in use for defence and construction, including axes, knives, saws and needles. A number of important machines had also been developed, such as the energy-storing bow for shooting arrows, the levered spear-thrower, and the wheeled cart for transport. Advances from hunting into agriculture led to the invention not only of special tools for working the soil but also of carpenters'

This section is an extract from the author's Keynote Address to the Fifth World Congress of the International Federation for the Theory of Machines and Mechanisms held in Montreal in 1979. The proceedings of the Congress were published by the American Society of Mechanical Engineers.

tools for making them. By about 3000 BC, when important developments in metal-working took place, highly structured states carried out major construction projects, often involving thousands of men at a single site. They achieved accuracies that belied the crudeness of their machines; the greatest deviation from a right angle at a corner of the Great Pyramid of Cheops is about 0.05° and the maximum difference in length of a side is about 1%.

A period of sustained progress then drew to an end. Advances in practice were slow until the flourishing of Greek civilization 1000 years later, when the widespread use of cheap iron led to another era of progress. Another major contribution was also introduced about that time. Scientists, of whom the greatest was Archimedes, took an interest in mechanical matters, despite the widespread disfavour in which the mechanical arts were held by leading thinkers of the day. As well as discovering scientific laws, Archimedes is credited with the invention of the compound pulley, the screw-pump that bears his name, and a variety of military weapons. Over a century later, Hero of Alexandria wrote several treatises on machines and mechanisms then in use, including the first known reference to a gear train, but throughout the period of the Roman Empire the most notable advances, except for materials of civil and military construction, came from the existing technology rather than from renewed invention.

The turbulence that followed the collapse of the Roman Empire led to the beginnings of a new machine-based civilization in Europe. The ideas of other societies were absorbed, mixed with indigenous mechanical advances, and multiplied many-fold in practice. Europe was by no means the most technologically advanced society of the times – one of the enduring puzzles of history, for example, is the migration of earlier mechanical invention in China – but the Europeans were forced to change their ways as the structures of their society changed.

They seized, for example, on the water-wheel, the windmill and improved harnesses for animals as major contributions to the generation of power, developed escapement-controlled mechanical clocks, invented the lathe, and produced the first recorded use of mechanisms for converting reciprocating into circular motion via a connecting-rod. These and other contemporary inventions were the advance notices of a technological age. Moreover, their success led to a new appreciation not only of mechanical cause and effect but also of the ability of man to design rationally. Thus the 13th century monk and scientist Roger Bacon wrote: 'I will tell first, therefore, of the wonderful works of Art and Nature, in order to assign to them afterwards their causes and means; in these there is nothing of a magical nature.' The role of monasteries, not only as centres of religion and learning but also as sources of orderly mechanical measurement of time, has been well marked by historians.

The ensuing era of the Renaissance led to the first sustained development of the science of mechanics since the statics of the Greeks. It also generated a major thrust forward in mechanical invention. The universal genius of Leonardo da Vinci (1452–1519) perceived the significance of relating the two aspects of theory and practice. Not only was he arguably the most prolific inventor of all time but he also investigated fundamental physical laws covering an immense field of interest, including physiology and bioengineering, optics, fluid dynamics and the mechanics of machines. His famous note-book of over 5000 pages of closely written text and diagrams provides a happy hunting-ground for interpreters to this day. Some examples are shown in Fig..1.3.

Fig. 1.3 *Leonardo da Vinci and a few of his inventions.*
(a) *Self-portrait;* (b) *self-centring thrust bearing;* (c) *winch;*
(d) *parachute;* (e) *armoured car.*

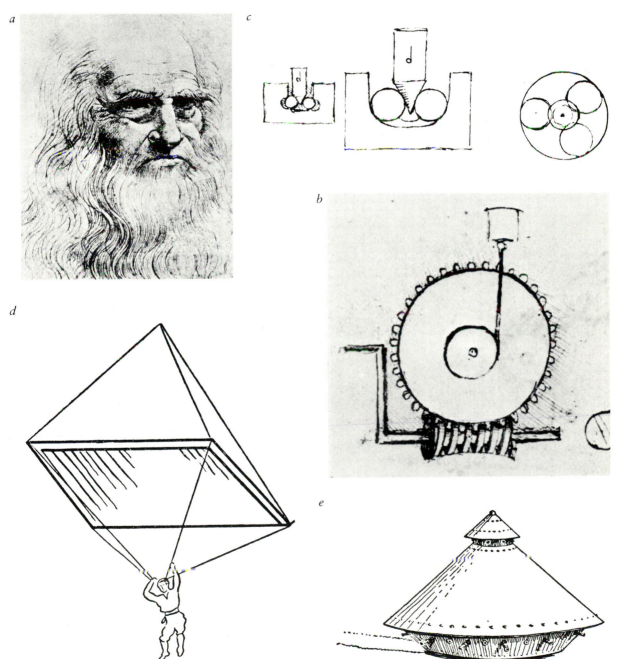

It was still too early, however, for scientific theory to exert a major influence on practice, which remained largely empirical. A vast expansion in international trade provided fertile conditions for the inventor, notably in textile machinery, mining, metallurgy and machines for the generation of power. Moreover, the foundations of industrial society were being laid by the concentration of production in factory organizations employing the newly invented machines and concentrated work-forces. Naturally, such developments attracted the hostility of some craft guilds. The record of the times is punctuated by various measures, some spontaneous and other legislative, designed to protect the interests of the *status quo*.

The 17th and 18th centuries saw the major changes in social structures that provided the basis of the Industrial Revolution. Manufacturers were not only identifiable as a major force to be reckoned with, but they also had a direct interest in encouraging the technical progress upon which their prosperity depended. Within about 200 years, the state of the industrialized countries has been transformed by the ways in which man used his new machines. For example, Newcomen's first steam engines (drawing on Toricelli's discovery of atmospheric pressure) were in use early in the 18th century in Austria, Belgium, France, Germany, Hungary and Sweden as well as in England; there were over four million spinning spindles in Arkwright's frames in Great Britain alone by the same period; the Jacquard weaving loom appeared in 1804; by 1800, Watt's steam engines were widely in use in coal-mines, cotton-mills, foundries and forges; in the USA, Fulton developed steam engines for the propulsion of ships in the early 1800s (it is alleged that Napoleon asked the French Academy of Sciences for a report on this matter, but, with an independence characteristic of academic institutions, they declined); and Stephenson's first passenger-carrying steam locomotive ran in 1825. By now the technological age was well and truly launched.

Theory took some time to catch up. During the 14th century, the idea of instantaneous velocity had been mooted by a group of scholars at Oxford. In France, Oresme concluded that it could be represented as a function of time on a two-dimensional graph. Buridan, rector of the University of Paris, called into question the Aristotlean idea that a projectile was continuously being pushed forward by the air it displaced, conceiving an intrinsic quantity called impetus, akin to momentum, as an alternative hypothesis.

These and other discoveries marked the opening stages of a revolutionary change in our understanding of the natural world. They were absorbed into the mechanics of Galileo and Newton, but much time had yet to pass before this ferment of intellectual theory influenced the empirical development of machines. Thus, Kepler, a geometer and mystic, described celestial motions by mathematical formulae; Galileo, the father of observational astronomy, investigated terrestial mechanics and foreshadowed Newton with the idea that a continuing force caused a continuing *change* in the velocity of a moving body. However, it was reserved for Newton to establish the law of gravitation and to affirm the laws of motion that form the general basis of dynamics. The publication in 1687 of what has been described as the greatest of all scientific works, *Principia Mathematica*, opened the door for a vast expansion of dynamical theory. Fig. 1.4 shows the author and the frontispiece of the book.

a

b
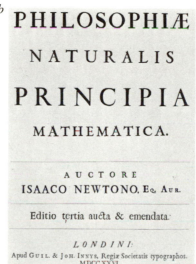

Fig. 1.4 *Sir Isaac Newton and frontispiece of* Principia Mathematica.

The 18th century was a period of intense activity. Euler, followed by d'Alembert, introduced the concept of inertia forces. He also laid the foundations of gyroscopic theory (although the word gyroscope was not coined until the physicist Foucault realized the practical potential of the instrument almost a century later) and extended the extremum principles of Fermat and Maupertius by mathematical proof. After doing so, he added the illuminating remark: 'The strength of this argument may not be sufficiently clear . . . I leave the task to others who make a profession of metaphysics.' A few years later, Lagrange quickly removed the professional metaphysicians from the field of debate; moreover, his *Mechanique Analytique* (1788), as indicated by its title and a total absence of diagrams, presented a mathematical point of view that has been the delight of many readers to the present day. It also reflected a change in emphasis from fundamental physical laws to procedures for the analysis of systems of continuous and connected bodies.

The foundations of mechanics were being explored against a background of an enormous development of industrial machines. Many of the leading scientists of the time made major contributions to the systematic study of machines and mechanisms as a branch of general mechanics. Indeed, the way in which these outstanding figures were called upon to consider practical problems is often overlooked. For example, a committee of the French Academy of Sciences including d'Alembert, Legendre and Monge was invited 'to examine means of improving navigation in the Realm', in the course of which they reported not only on fluid resistance to ships' motions but also reflected widely on experimental methods. Coriolis was concerned

with the motion of water-wheels when he developed his famous
theorem concerning moving coordinate frames.

This relationship between theory and practice began to attract
serious attention in the early part of the 19th century. In order to
glimpse the spirit of the times, it is helpful to appreciate the vigour of
the debate inspired by Ampère in 1830 when he laid down limits of
the science of mechanisms. 'It,' he considered, 'must therefore not
define a machine, as has usually been done, as an instrument by the
help of which the direction and intensity of a given *force* can be
altered, but as an instrument by the help of which the direction and
velocity of a given *motion* can be altered.' Force and its effect were to
be completely excluded. Further 'To this science . . . I have given the
name Kinematics, from $\kappa\iota\nu\eta\mu\alpha$, motion.'

Ampère's proposal was taken up enthusiastically by Willis,
Professor of Natural and Experimental Philosophy at Cambridge,
who in 1841, after remarking that mechanics had 'occupied the
attention of nearly every mathematician of eminence who has arisen
in the world', went on to note that but few had concerned themselves
with mechanisms in the sense of Ampère. Indeed he seems to have had
some reservations about Ampère himself, for he added 'It is much to
be regretted that this distinguished writer did not attempt to follow up
this clear and able view of the subject, by actually developing the
science in question'. Throughout his classic work, however, Willis
retained a strict interpretation of the purity of the subject, resolutely
excluding considerations not only of force but also of non-rigid media
such as water and air. Like earlier writers of the period he was
preoccupied with the general taxonomy of mechanisms, but he also
offered a sharp insight into what we now term kinematic design,
including change-gears, continuously variable speed transmissions
and path generation.

These matters had also been considered in Russia, notably by
Chebyshev. As well as establishing basic theories of kinematic
synthesis he saw the importance of drawing together the threads of
theory and practice for 'it is not only practice which benefits from
(rapprochement); science itself develops under its influence'. The same
theme was given elegant expression by Reuleaux in his oft-quoted
classic *The Kinematics of Machinery*, translated into English in 1876
by Kennedy. To the interested reader his Introduction may be warmly
commended:

> *'Theory and Practice,' he wrote, 'are not antagonistic, as is so often
> tacitly assumed. Theory is not necessarily unpractical, nor Practice
> unscientific, although both of these things may occur . . . It is not a
> matter of merely setting forth in a new form and order that which is
> already known . . . On the contrary, if the new theory is to lay
> claim to general interest, it must be capable of producing something
> new; it must make problems solvable which before could not be
> solved in a systematic way.'*

A sense of direction

One of the key concepts in understanding motion is that of vectors. It is all a matter of geometry. Suppose, for example, that a point moves from *A* to *B*, Fig. 1.5. On a separate diagram, we can represent its actual displacement *AB* by a directed line segment or arrow (*ab*) drawn in the same direction as *AB* to a chosen scale. If the point in question moves to another position *C*, we can draw another arrow (*bc*) to the same scale to represent the second displacement *BC*. It is

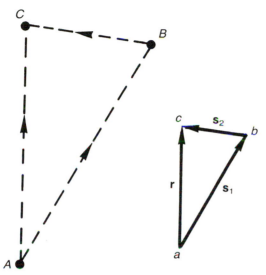

Fig. 1.5 *If a point is displaced from* A *and then from* B *to* C, *its resultant displacement is* AC. *If the displacements* AB *and* BC *are represented by vectors (or arrows)* ab *and* bc, *respectively, the resultant displacement* AC *is represented by the vector* ac.

easy to see that the overall or resultant displacement *AC* is correctly represented by the arrow (*ac*). Moreover, we can write

$$(AB) + (BC) = (AC)$$

and

$$(ab) + (bc) = (ac),$$

or, in vector notation,

$$\mathbf{s}_1 + \mathbf{s}_2 = \mathbf{r},$$

where bold type indicates a vector and ordinary type its length.

Once the principle is established for two successive displacements, it can be extended to any number, so that

$$\mathbf{s}_1 + \mathbf{s}_2 + \mathbf{s}_3 + \ldots = \mathbf{r}.$$

The *vectors* in the discussion are the arrows such as (*ab*), and the displacements which they represent are termed vector quantities. Certain conditions must be fulfilled before a candidate can qualify as a vector quantity: it must have a magnitude and a direction, and it must combine according to the simple geometrical rule stated above. As we have seen, displacement of a point qualifies without difficulty. The velocity of a point also qualifies. Note, however, that in combining two or more velocities, the constituent parts exist *simultaneously*, as when the dreamer on the beach gets up and walks on the moving

Fig. 1.6 *Angular velocities can be represented by vectors. If the aircraft rolls with angular velocity* **r** *and pitches with angular velocity* **p**, *its resultant angular velocity is represented by the vector* **q**.

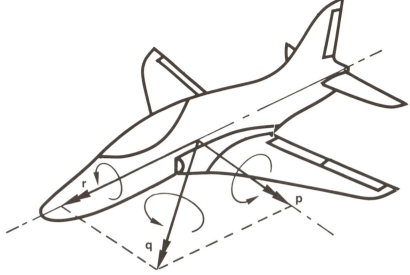

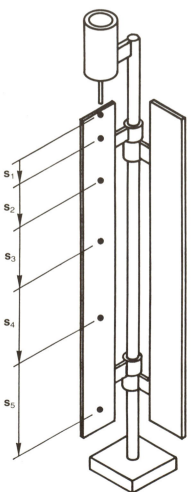

Fig. 1.7 *Leonardo's measurement of* **g** *with falling drops.*

surface of the Earth. Note also that *speed* is simply the *magnitude* of a velocity; it is not by itself a vector quantity.

Other familiar vector quantities include acceleration of a point, linear momentum of a particle or a body, and also certain quantities associated with angular motion, such as *angular* velocity of a line or a rigid body. Fig. 1.6 illustrates the latter. If the aircraft is rolling about its fore-and-aft axis, its angular velocity can be represented by a vector **r** (preferably double-headed to distinguish it from vectors representing linear quantities) pointing along the fore-and-aft axis of roll. If the aircraft is also pitching about a transverse axis (nose down, tail up), its angular velocity of pitch can be represented by a similar double-headed vector **p** about the transverse axis of pitch. The *resultant* angular velocity of the aircraft is represented by the vector sum **r** + **p** = **q**.

Curiously, although angular *velocities* are vector quantities, angular *displacements* are not. It is true that angular displacements have both magnitude and direction, and so it is tempting to represent them by vectors. However, they fail on the third test of combination, as may be confirmed by rotating this book through successive rotations of 180° about each of three perpendicular axes fixed in space. You will find the book comes back to its original position – the resultant physical angular displacement is zero. Since the sum of three perpendicular vectors is clearly *not* zero, we may safely conclude that angular displacements are not vectors. They *can* be represented by complicated mathematical means beyond the scope of this book, but fortunately there is a redeeming feature to hand. When the rotations are very small, they are *almost* vectors, and in the mathematical limit of infinitesimal rotations – which determine angular velocities – they become vectors exactly.

Leonardo da Vinci carried out the experiment shown in Fig. 1.7 to find out how fast things fell down. We can interpret the results in terms of velocities and acceleration vectors – and, incidentally, get a good approximation of the value of **g**, the acceleration caused by the Earth's gravity. Small drops of water are allowed to fall at a uniform rate of, say, ten per second from a narrow glass tube. When a regular pattern is going well, the two 2 m long hinged boards are suddenly clapped together, capturing the instantaneous positions of about six drops. As the distances between successive marks show the progressively larger displacements, s_1, s_2 . . . travelled in equal intervals of time by a *single* drop, it is an easy matter to work out its average velocity during each interval. We simply divide the measured distance by the interval of time T. Such a set of vectors v_1, v_2 . . ., all pointing downwards, allows us to work out the average acceleration **a**: all that is necessary is to measure the difference in successive velocities and once again to divide by the time T. Admittedly, the acceleration **a** will not be quite the same as **g**, because the air through which the drops fall slows them down, but at the low speeds involved the effect is small.

In the experiment, the drop rate was set at ten per second: three velocity vectors were represented by cutting off thin metal rods to the distances between successive drops, and two acceleration vectors were represented in the same way by cutting off lengths equal to the differences between the velocity vectors. The average acceleration was found to be 9.5 m/s². The correct value of **g** in London is 9.8118520 m/s². All things considered, not too bad a result.

In Leonardo's experiment, everything happened in the same direction. When the *direction* of a velocity changes, however, the acceleration will have a direction that differs from that of the velocity. Even if the speed of the bob at the end of the string in Fig. 1.8 is constant, it will still have an acceleration because its velocity is constantly changing direction.

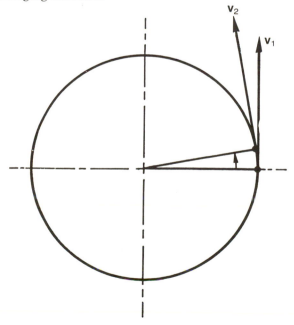

Fig. 1.8 *Centripetal acceleration of a point travelling around a circle at constant speed.*

Change in velocity

Remembering that the acceleration of a point is simply the rate of change of its velocity, we can see from the figure that the change of the bob's velocity is a vector that points towards the centre of the circle. The instantaneous acceleration can be found by dividing the change in velocity by the time interval involved and working out the result as the interval becomes progressively smaller. This particular acceleration occurs so often in machines that it is given a special name, *centripetal* acceleration (after the Latin *petere*, to seek), and it can be very large and very dangerous.

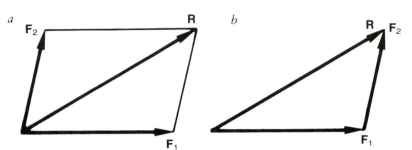

Fig. 1.9 *Combining forces.*
(a) *The famous parallelogram,*
and (b) *the alternative triangle.*

Another true vector quantity is force. A force has both magnitude and direction. Moreover, when two forces act a point, experiment shows that their physical effect can be represented by a resultant force determined by the rule of vector addition. The geometrical construction in this case is usually known as the parallelogram of forces, Fig. 1.9(*a*), but the triangular diagram, Fig. 1.9(*b*), will do just as well, as long as we remember that the forces actually act at the same point. The exact correspondence between the rules for adding displacement on the one hand and forces on the other is, to say the least, striking. On further investigation, it reveals a fundamental aspect of the physical world, for if the rules of addition differed we should have an unlimited source of energy at our disposal.

An interesting result of the vector addition for forces can be demonstrated with the help of a piece of rope. First, consider the forces acting on a nail hammered in a wall and supporting a hanging picture, Fig. 1.10(*a*). If everything is in equilibrium, the upwards force exerted on the picture cord by the nail must be equal to the weight of the picture, **W**. Next, consider the equilibrium of a small piece of cord in contact with the nail, as shown in Fig. 10(*b*). Since it, too, is in equilibrium, the two tensions **T** and the force **W** acting upwards on the cord must balance exactly, i.e. the resultant of the two **T**'s is precisely equal and opposite to the **W** upwards. Fig. 10(*c*) shows that if the inclined segments of the cord are close to the horizontal, the tensions **T** will be much larger than the weight **W**: if the cord were to remain horizontal, its tension would be infinite – but it would of course have snapped long before then. This suggests the following tug-of-war experiment. Tie one end of a strong rope to a fixture. Invite several heavy experimentalists to pull on the other end, telling them that you intend to pull them forward, possibly with one hand. They will not believe you. Grip the middle of the rope firmly and pull it *sideways*. With luck, your pull **W** will be more than enough to

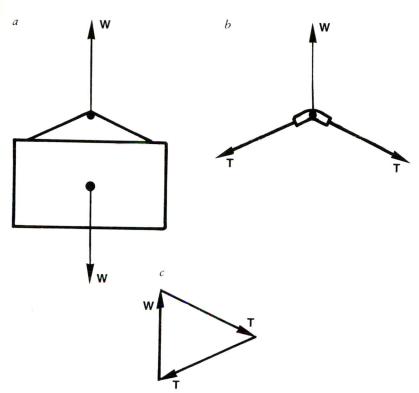

Fig. 1.10 *The hanging picture. In (a), the nail holds it up with a force* **W**; *in (b), the two equal tensions in the cord must balance* **W**, *as shown in (c), the triangle of forces.*

overcome the resultant of the tensions **T** in the piece of rope in your hand, after which you can explain the theory of vectors to your collaborators. Although in the actual trial one end of the rope was firmly tied to a railing, three determined (and weighty) young men at the other end, all wearing rubber-soled shoes, proved too much for the young lady pulling sideways on the middle of the rope, and the lecturer had to lend a hand.

Another simple construction is the reverse of combining vectors. If we start with a single force **R**, for example, we often wish to replace it by two equivalent forces **F**$_1$ and **F**$_2$ (Fig. 1.9(*a*) again), which are termed the *components* of **R**. When **F**$_1$ and **F**$_2$ are perpendicular to each other, they are especially useful in analysis and are called the resolved parts or Cartesian components of **R**, in honour of the great mathematician Descartes.

A matter of forces

It is an astonishing fact that outside the nuclei of atoms only two kinds of forces are known to exist in the entire universe. One is gravity and the other is electric. All other forces, whether from solid, fluid, gaseous or magnetic sources, are derived from one or the other. Physicists using elaborate equipment are at present carrying out large-scale and exciting experiments to determine the relations between them and the mysterious forces acting across tiny distances of about 10^{-15} m (one-millionth of one-millionth of a millimetre) inside the nucleus. Their eventual aim is to bring all forces including gravity into a 'grand unified theory' – a challenge on a cosmic scale.

Gravity is so familiar that we can easily forget it is also very mysterious. Each body in the universe exerts a gravitational force on every other body. Such forces are always attractive, i.e. they tend to pull the bodies together. The further apart the bodies are, the smaller the forces. If the distance is doubled, the force is reduced to one-quarter of its original value; if the distance is multiplied by ten, the force becomes one-hundredth, i.e. the force varies inversely as the square of the distance. The attractive force also depends on the masses of the bodies, and its magnitude can be expressed mathematically as

$$F = G(m_1 m_2 / R^2),$$

where m_1 and m_2 are the two masses concerned, R is the distance between their centres and G is a universal constant. Newton discovered this fundamental law at the age of 23. Note that the law implies that the force exerted by A on B is equal and opposite to the force exerted by B on A, a result that is true not only for gravity but for all other forces as well, as Newton discovered even earlier.

The value of the universal constant G is 6.670×10^{-11} newton metres^2/ kilogram^2 (Nm^2/kg^2). So the force of attraction between two bodies 1 m apart and each of mass 1 kg is 6.670×10^{-11} N, about one-hundredth of one-thousandth of the weight of a dandelion seed. Only when at least one of the bodies concerned has a very large mass, such as the Earth (5.98×10^{24} kg), is the gravitational force perceptible by ordinary means; when we weigh something, the weight is the gravitational pull of the Earth. Nevertheless, gravity is all pervasive and acts on all mass throughout all space – without, as yet, any known explanation. Physicists describe such a phenomenon as a *field* of force, filling all space and possessing both strength and direction. Almost 200 years after Newton, Einstein put forward proposals dealing with small deviations observed in the motion of the planet Mercury that could not be explained by existing theory. It represented that the properties of time and space are themselves changed, or warped, by the presence of big stars. In view of the complicated nature of the new theory, it is fortunate that the older theory works so well for our usual purposes.

The other kind of force depends on electric charges. It has certain resemblances to gravity and also certain marked differences. The *electrostatic* force exerted by one stationary charged body upon another acts along the line between them with a magnitude expressed mathematically as

$$F = k(q_1 q_2 / R^2),$$

where q_1 and q_2 are the measures of the charges, and R is the separation; if q_1 and q_2 are measured in coulombs, R in metres and F in newtons, the constant k has the value 9.00×10^9 N m^2/C^2. This result looks very much like the expression for gravitational force, but whereas mass in the latter is always positive the electric charge may be positive or negative; correspondingly, the electrostatic force will be repulsive or attractive according to whether the charges have the same or different signs.

Moreover, physicists have discovered that all nature possesses a basic elementary charge, that of a single electron inside an atom. Its value is $e = 1.60 \times 10^{-19}$ C, and by convention it is given a negative sign. In an ordinary atom, made up of a number of electrons moving in orbit around a central nucleus at distances of about 5×10^{-11} m, the negative charges of the electrons are matched exactly by the equal positive charges of the same number of protons in the nucleus, and because the other constituent of the nucleus, the neutron, has zero charge, the net charge of the whole assembly is precisely zero. This marks another striking difference between electric and gravitational forces, for whereas the electric force field created by a large number of atoms in a body is zero (except in special cases) the gravitational field is never zero.

One special case is very familiar. If a balloon is rubbed by a cloth, some of the electrons are rubbed off, leaving the balloon with a net positive charge. As a result, it will attract negative charges from a nearby surface and, given the opportunity, the balloon will stick to the surface by courtesy of the mutual attraction between unlike charges. Fig. 1.11 shows a volunteer wearing a balloon on her head with the

Fig. 1.11 *A case of electrostatic attraction.*

help of electrostatic forces. Similar effects occur when you comb your hair, or rub a glass rod with silk. The crackling that can sometimes be heard, and the sparks that can sometimes be seen, show that the air between has been converted by a strong electric field from being an insulator, which resists the passage of electrons, into a conductor, which allows electrons to move freely. Atmospheric lightning is the same thing on a more impressive scale.

It may seem strange, but electric forces account for much of our ordinary experience of pushes and pulls. It becomes less strange when we remember that all materials are made up of atoms, like the array shown in Fig. 1.12, and that each atom consists of a positively charged nucleus and negatively charged electrons. Any attempt to rearrange the structure is resisted by electric forces between neighbouring nuclei and the electrons of adjacent atoms. These forces control the deformation of the material and they can be very strong, as anyone knows who has tried to indent the surface of a piece of steel.

Liquids and gases, of course, behave differently. The internal forces

Fig. 1.12 *Atomic structure of diamond.*

holding atoms together in a liquid are very much smaller than those in a solid, for reasons not yet fully understood. In gases, the individual molecules or atoms are so far apart that their internal electric forces are even less, so that changes in volume as well as in the shape of a mass of gas can be achieved very easily by external forces. And the pressure of a gas on the walls of a container can be calculated very well from the mass of the molecules and the number and the speed of the bumps with which they strike the surfaces involved.

The mighty electron and its machines

Within the last 150 years or so, scientists and engineers have learned how to make use of the motion of tiny electrons to create a vast family of electrical machines. Electric current in a wire is a good starting point. When the positive and negative terminals of a battery are connected by a conducting wire, an electric field is generated along the entire length of the wire. Inside the wire, a certain number of electrons that ordinarily are bouncing around between atoms are persuaded by the field to move in one direction only. Since their charges are all negative, they drift from the negative end to the positive end. Such a flow of electrons constitutes an electric current. Conventionally, the current is considered as flowing from positive to negative, i.e. in the opposite direction from the flow of electrons. There is another peculiarity, too, this time a real physical one. As soon as the terminals are connected, electrons start to move practically instantaneously at all points along the wire, but their average drift velocity is actually very small, less than one millimetre per second for a current of one ampere (one coulomb of charge passing a given point per second).

A remarkable property of electric currents is that the wires carrying the currents exert forces on each other. If currents flow along two parallel wires, the wires will attract or repel each other, according to whether the currents are in the same or the opposite direction. If the wires are not parallel, similar effects occur, although the direction and the strength of the forces depend on the particular arrangement. These forces cannot be electrostatic, as the net charge in each wire is always zero. Instead, they reveal a different kind of force altogether, one associated with the *motion* of electrons. This astonishing discovery was made soon after the Danish physicist Oersted had found in 1820 that, when he passed an electric current through a wire placed near and parallel to an ordinary magnetic compass needle, the needle swung away from north. The current had evidently created a new force to which the needle had responded, and it began to look as if that force was similar to the magnetic forces already familiar in nature. So it proved, and the relationships between electricity and magnetism began to be unravelled.

The picture we now have is that of a magnetic field of force created by a current and behaving in the same way as a magnetic field associated with a natural magnet. Like a gravitational field, the magnetic variety has both strength and direction at each and every point in space, as may be demonstrated by the way a compass needle

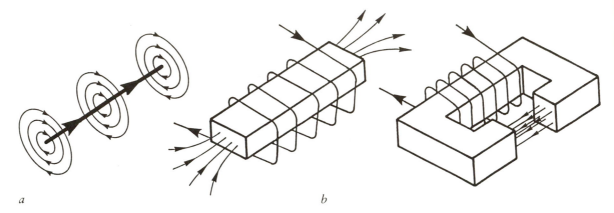

Fig. 1.13 (a) *Magnetic field
around a straight conductor, and
(b) electromagnets.*

responds at that point. Fig. 1.13 (*a*) shows the circular lines of force
marking the field directions near a straight wire. If the wire is
arranged as a coil, the field inside the coil is directed along the axis of
the coil and may be concentrated by a bar of iron placed inside the
coil, as in the electromagnets shown in Fig. 1.13(*b*).

Natural magnetism, including that of the Earth itself, had been
known and used for centuries before Oersted made his discovery. It
was thought to be associated with a particular kind of iron ore, now
called magnetite, but the explanation remained a mystery. Now we
know that even natural magnetism is caused by electric currents. The
spinning of electrons around the nucleus of an atom acts like an
electric current in a tiny coil. In most elements the axes of spin are
arranged in a random way, and their magnetic fields cancel each other
out, but in some elements they form small domains having definite
magnetic fields. These domains can be made to line up, and in some
elements even to stay lined up when the cause has been removed.
These are the permanent magnets. In the case of the Earth, the
explanation depends on electric currents circulating in the molten core
that are believed to be associated with the spinning of the Earth
around its polar axis.

The chapter of discovery that began in 1820 soon led to other
unexpected phenomena of immense practical importance. If, by
means of their magnetic fields, electric currents flowing along wires
result in forces between the wires, thereby causing the wires to move,
could things happen the other way around as well, i.e. could the
motion of a conducting wire through a magnetic field generate its own
electric current, as shown in Fig. 1.14? The answer was yes. And so
electricity was produced from a magnetic field.

And that was by no means all. Michael Faraday discovered that if
instead of moving the wire through a stationary magnetic field (or
moving a magnetic field through a stationary wire) he varied the
strength of a magnetic field passing through a *stationary* wire, once
again an electric current was generated. This enabled him to generate
electricity without actually moving the physical parts at all. For if the
magnetic field is provided by an electromagnet as in Fig. 1.14, the field
could be varied simply by varying the strength of the primary current.
The result is that a new, secondary current is induced in the stationary

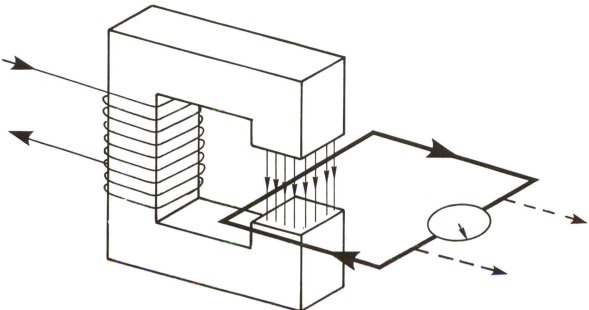

wire. Faraday developed his ideas on this and related aspects of electromagnetism in a series of classical experiments that laid the foundations for the enormous range of electrical machines on which we now depend. Let us look first at a simple type of electric generator.

Fig. 1.15(a) shows a model similar to that used by Faraday. When the coil of wire is rotated in the permanent magnetic field, a current is induced in both sides of the coil flowing in the direction ABCD. It can be registered by means of an instrument, an ammeter, placed in the circuit. Its strength depends on the rate at which the wire travels across the field and will be a maximum when the rate is greatest, i.e. in the position shown. Half a revolution later, when the two parts of the coil have exchanged their positions, the current flows the other way.

Fig. 1.14 *Electromagnetic induction. An electric current is induced in the closed coil of wire when it is pulled through the magnetic field.*

Fig. 1.15 *The generation of electricity. (a) ac generation, and (b) dc generation via a split ring.*

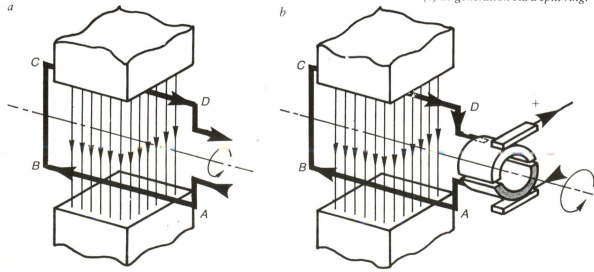

Such an arrangement generates an alternating current. If, on the other hand, the rotating ends of the coil make their contacts to the external circuit via a split ring, called a commutator, such that the upper side of the coil – whichever it happens to be at the time – always makes contact with the same external connection, then the current will continue to flow in the same direction into the outside circuit although it alternates in the coil itself. The result is a direct current machine, although in our simple arrangement the current would vary in strength according to the position of the coil. Things would clearly be improved for both a.c. and d.c. machines if, instead of a simple magnetic field created by a single north–south pole pair, a larger number of pole pairs were arranged around the axis of rotation. Moreover, if these fields were created by powerful electromagnets, and if, instead of a coil made up of a single diametral loop, several loops were wound around a substantial rotor, as shown in Fig. 1.16, the result would be even more effective.

Naturally, something would have to drive the coils around in order to supply the mechanical energy that emerges from the machine in electrical form. In the large machines that generate electricity for the national grid, the driving torque is supplied by giant steam turbines. And instead of an electromagnet in the casing of the machine generating electricity in the rotating coils, the arrangement is reversed: the rotor coils, supplied with direct current via special contacts, form a rotating electromagnet which induces the electrical output in coils fixed in longitudinal slots in the stator. The reason is simple: the direct

Fig. 1.16 *A multi-pole generator.*

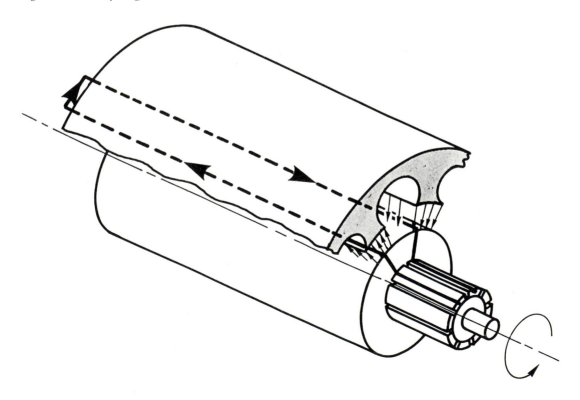

current supplied to the rotor, usually by means of a small generator
directly coupled to the rotor-shaft, is much smaller and of lower
voltage than the output of the generating coils, and so the problems of
transmission across moving contacts are much reduced. The details of
the winding, coil connections, insulation and the many difficult
mechanical problems including stressing, heat transfer, vibration and
control call for engineering skills of a high order.

Electric motors work on the same principles but apply them the
other way around: they are supplied with electrical energy and
transform it into mechanical energy, usually in the form of a driving
torque at an output shaft. The simplest construction can be illustrated
by means of Fig. 1.17. Suppose an electric current is passed through
the coil. As a result of the interaction with the magnetic field, forces
are generated on the coil, to the left on the upper section *AB* and to
the right on the lower section *CD*. If the coil is free to rotate, it will be
driven around by these forces, the resultant torque being strongest in
the position shown but becoming zero when the coil is horizontal. As
the coil overshoots the horizontal position, the direction of the current
in it is reversed by means of the split ring through which the current is
supplied. The upper section of the coil (now *DC*) still generates a
force to the left, and the lower section (now *BA*) to the right, and so
the motion continues in the same direction. Such an arrangement
forms the basis of the simplest form of motor, to which electricity is
supplied in the form of direct current. As in the case of a generator,
better results can be achieved by using electromagnets instead of
permanent magnets, by providing more poles arranged around the
circumference of the casing, and by wrapping a large number of coils
around a solid rotor.

The relation between the current supplied to the field
electromagnets on the one hand and to the rotor coils or armature on
the other determines the driving characteristics of the machine. If, for
example, *alternating* current is fed to the coil in Fig. 1.17 instead of
direct current, and if an alternating current precisely in step with the
coil current is used to generate the magnetic field of the magnet, then

Fig. 1.17 *A dc motor. In the
position shown in (a), the current
supplied to the coil flows in the
direction* ABCD, *and the forces
exerted on the segments* AB *and*
CD *are in the directions shown.
In (b), half a revolution later, the
current flows in the direction*
DCBA, *but the forces continue to
turn the coil in the same direction.*

a   *b*

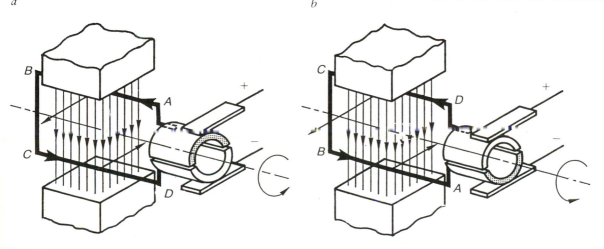

the resulting forces on the coil will have the same direction as if direct
current had been used, and we have an a.c. motor.

The range of electric motors is very wide, extending from small
machines such as those used in food mixers and refrigerators, to
medium sized machines for lifts and machine tools, and to large
machines for traction and other industrial uses. Their efficiency in
converting energy from electrical to mechanical forms is very high;
although they work best at high speeds of rotation, lower speeds can
readily be achieved by means of reduction gearing at the output
shaft.

An important type of a.c. motor depends on the induction of
electricity in conductors when a magnetic field moves past and
through them. In a *rotational* induction motor, alternating current is
fed to electromagnets spaced out around a cylindrical stator in such a
way that the peak current in each pole pair is reached at successive
times: the effect is a magnetic field that rotates around the cylinder.
Suppose that inside the cylinder we place a rotor with bar-like
conductors laid out in longitudinal slots distributed all around the
periphery and connected at their ends, as in Fig. 1.18; no external
current is supplied to these conductors and no external electrical
connections need be made. As the stator's field turns around, like an
invisible circular brush with its bristles pointing radially inwards, it
induces current in the rotor's conductors according to the principles
of induction discovered by Faraday. The rotor current interacts with
the spinning magnetic field to generate forces on both stator and
rotor: the stator cannot move but, provided the rotor is free to rotate,
it will spin around in pursuit of the rotating field.

Similar principles are involved in a *linear* induction motor. Imagine,
as in Fig. 1.19, that the cylinder forming the stator of a rotating motor

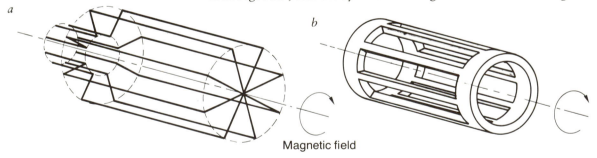

Magnetic field

Fig. 1.18 *A rotational induction
motor. When suitably phased
alternating currents are supplied
to the independent coils in* (a), *a
rotating magnetic field is created
in the interior space. If the slotted
rotor* (b) *is placed inside the coils,
electric currents are induced along
its bars which interact with the
magnetic fields to turn the rotor.*

Fig. 1.19 *A linear induction
motor.*

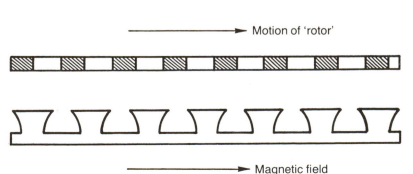

Motion of 'rotor'

Magnetic field

is unrolled and laid flat, and that the rotor is replaced by an object to be driven forward. An alternating current is supplied to the stator's electromagnet in such a way that a magnetic field travels from one pole to the next and so on down the line. Electric currents are thereby induced in a conductor placed across the direction of motion of the field, and, as in the rotational machine, the resulting forces compel the conductor to follow the field, this time in a straight line.

The effect was demonstrated with the help of a small linear induction motor kindly loaned by Professor Eric Laithwaite. Its stator consisted of a set of electromagnets supplied with alternating current so as to provide the travelling field, and the rotor was simply a flat sheet of aluminium. When the current was switched on, the rotor was driven rapidly along the stator and shot off the end, fortunately without causing any damage.

This behaviour suggests much bigger possibilities. Provided the parts an be made big and strong enough, the rotor could carry people around – perhaps as a train. Naturally, the stator would have to be extended – at least several miles for a decent sort of railway – and because the stator has to be supplied with alternating current along its length, the original arrangement may not be very practical. There is, however, another way of doing it. According to Newton's third law of action and reaction, the force on the stator is equal and opposite to the force on the rotor. Suppose the rôles of these two parts are exchanged, i.e. we make a short 'stator' and a long 'rotor', then fix the 'rotor' to the ground and allow the 'stator' to move. The terminology gets rather confused, but we finish up with a far more sensible arrangement – a short travelling vehicle driven by external current, and a long simple track. And if independent electromagnets are used to suspend the vehicle over the rails, the train should be able to run without wheels.

British Rail has actually done all this. The Mag-Lev train shown in Fig. 1.20 links the airport and the UK National Exhibition Centre at Birmingham. It is propelled by two centrally mounted linear induction motors and is suspended by means of eight powerful electromagnets, two at each corner to provide lateral stability as well as vertical lift. The weight of the unladen train is almost 3 t and its length is 3.5 m. Although its speed is modest, as befits a short-run service, it can negotiate curves and gradients with considerable agility.

There is an interesting point about magnetic suspension. If you try to place a pin a short distance beneath a magnet so that the magnet's pull exactly balances the pin's weight, you will soon discover a serious difficulty. Although such a position undoubtedly exists, it is also totally unstable. The slightest deviation upwards will increase the magnetic force and the pin will finish up clamped to the pole face; if the initial deviation is downwards, the pin will simply fall away.

The large beach ball complete with a metal cap, beneath the electromagnet in Fig. 1.21, illustrated the difficulty by refusing to remain in the balanced position. If, however, the current in the electromagnet is adjusted so that when the air-gap (as measured by a simple detector made up of a light and a photoelectric cell) decreases

Fig. 1.20 *The Mag-Lev train.*
Courtesy of British Rail.

Fig. 1.21 *Controlled magnetic*
suspension of a beach-ball.
Courtesy of British Rail.

the current is reduced, then the magnetic pull is also reduced and the gap is restored; if the gap increases, the detector once again comes into action, this time increasing the current and so reducing the gap. In a trial run without control, the ball either fell or clamped itself to the magnet. With control, it floated stably a few millimetres below the pole faces.

Such an arrangement provides a stable and effective magnetic suspension. It acts like a mechanical spring, but has the advantage that its stiffness can be adjusted by means of current control to offer either soft or hard rides. Overall, the costs and advantages of this type of train have to be tested by experience, and its prospects are still being investigated.

Kinematics, or how it moves

Dynamics is usually divided into two parts, kinematics and kinetics. Kinematics deals with positions and their rates of change without reference to mass or force, and kinetics relates kinematics to forces. Consider that humble but efficient machine, the bicycle. If you wanted to know how far it moves forwards for one revolution of the pedals, you would consider its kinematics. If, on the other hand, you wanted to know how hard to push the pedals to give it a certain acceleration, you would be in the field of kinetics. And if you were to ask why you can stay up without falling over, you would be into kinetics with a vengeance.

Let us look at the kinematic of gears. They have been in use for so long that they often pass by without notice. Yet they are of great importance and are used in many shapes and sizes, from tiny wheels the size of a pinhead to giants weighing many tonnes. Usually they are used to change speed between one shaft and another in order to give the best performance at the output of the machine.

Fig. 1.22 shows two simple spur gear wheels. As they turn together, their teeth must engage properly, which means that the distance between successive teeth in one wheel must be exactly equal to the

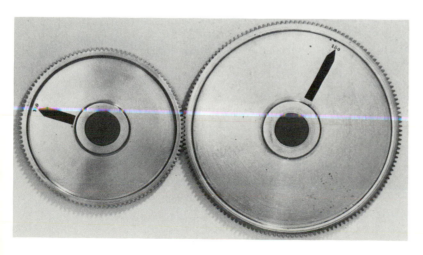

Fig. 1.22 *Spur gear-wheels. The smaller wheel turns 15/11 times as fast as the larger.*

corresponding distance in the other. For the same displacement at the circumference, the smaller wheel will rotate through the larger angle and hence will turn faster: to put it more exactly, the speed ratio is the *reciprocal* of the ratio of the radii or of the number of teeth.

One half of the audience was invited to count the revolutions of the small wheel, and the other half to do the same thing for the large wheel. At the start and stop positions the markers were in line and pointing towards each other. The wheels were turned slowly at first, then faster. Counters returned a score of 15 revolution for the small wheel and 11 for the large. The numbers of teeth (too small to be counted by the audience) were 110 for the small wheel and 150 for the large.

When gear wheels are arranged in combination, a large range of speed ratios can be achieved. Fig. 1.23 shows an epicyclic arrangement (an epicycle simply means one circle rolling on another). It consists of an outer wheel W with internal teeth, a central or sun wheel S, and three planet wheels P which engage with both W and S. The planet wheels are connected to each other at their centres by a circular ring R; each can turn on R about their centres and, in addition, R can also turn as a whole, carrying the planet wheels with it. There are three different ways in which the combination can move. In the first, the ring R is locked, Fig. 1.23 (*a*). When the sun is turned, it drives all three planets about their stationary centres and they in turn drive the outer wheel: from the details of the kinematics we can work out the speed ratio. If W is locked instead of R, an entirely different speed ratio is obtained between R and S, Fig. 1.23(*b*); if S is locked, yet another speed ratio is obtained between W and R, Fig. 1.23(*c*).

The numbers of teeth in W, P and S were 120, 40 and 40. With R locked, the speed ratio of W to S was $1/3$; with W locked, the speed ratio of R to S was $1/4$; with S locked, the speed ratio of W to R was $4/3$. Kinematics is the key.

In early days, gear teeth were simply short spokes sticking out from the circumference of the wheel. They did not give a very smooth transmission. In modern machines, which transmit large forces at high speeds, it is essential to have teeth that fulfil very demanding

Fig. 1.23 *An epicyclic gear. In (a), R is locked: when W is turned, the wheels rotate as shown to give a particular speed ratio between W and S. In (b), W is locked to give a different speed ration between R and S. In (c), S is locked to give a third speed ratio between W and R.*

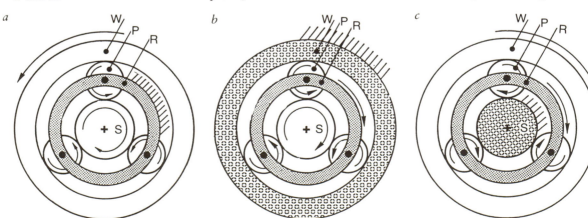

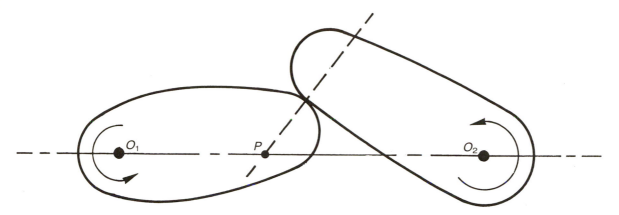

requirements. One important question concerns the precise shape of the teeth. If the shapes are wrong – like spokes for example – the driven wheel will move in a series of jumps, even if the driving wheel goes at a steady speed. If the shapes are right, however, the driven wheel can turn at a uniform speed, and the question for the designer is 'What shapes?' One answer is the involute shape, and to see how it works we first look at Fig. 1.24, where we imagine that one flat plate, cut out to a rounded sort of shape, is pushing another around. The point P, where a line drawn as shown through the point of contact cuts the line of centres, plays a crucial part in the proceedings, as the speed ratio depends on the ratio O_1P/O_2P. If that ratio stays constant as the bodies turn, i.e. if P remains fixed, we have a good transmission with a constant speed ratio. So now the question is 'What shape keeps P fixed?' That is a nice question of geometry, and Fig. 1.25 shows the answer.

Fig. 1.24 *Transmission of rotation: the speed ratio depends on the ratio* O_1P/O_2P.

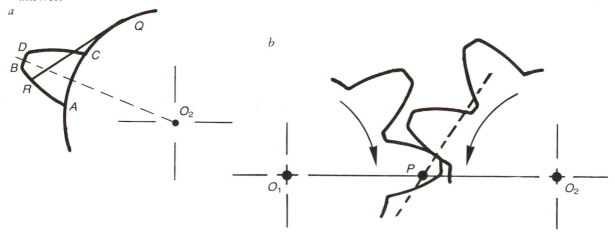

We start with a circle, and mark out the locus *AB* traced by the free end of a piece of string as it is kept taut and unwrapped from the circumference of the circle. That gives us one flank of the tooth. Another piece of string, unwrapped the other way, gives the other flank *CD*. When teeth of this shape engage with each other, it can be shown that the line through their moving point of contact always cuts the line of centres at the same point, and a steady speed ratio is

Fig. 1.25 (a) *Construction of an involute tooth profile.*
(b) *The point P remains fixed as the wheels rotate, giving a constant speed ratio.*

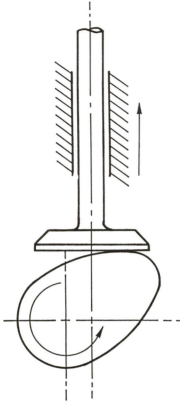

Fig. 1.26 *Cam and follower.*

assured. In practice, of course, the shapes are drawn out geometrically rather than with string, and there are many other factors involving loads, deformation, friction and so on to be considered as well. But the kinematic principles are the same.

An overhead projector showed the actual construction using string and transparent cut-outs for shapes. When two wheels of different colours were rotated in mesh, the point of contact between the teeth moved exactly down a fixed line towards the centre-line point *P*.

Another important engineering problem involves the changing of circular to straight-line motion, or the other way round. One method is to use a cam, as shown in Fig. 1.26. As the rounded member or cam rotates, it moves the follower forwards and backwards in a way that depends on the shape of the cam, so that a wide range of motion can be obtained by suitable design of that shape. The arrangement is often used when the follower has to be moved in perfect timing relative to the rotating cam, as is required in the valves of a reciprocating engine.

The principal parts of such an engine also transform the straight-line or linear motion of the pistons into circular motion of the crank-shaft. The arrangement, as shown in Fig. 1.27, consists of a connecting-rod *AB* and crank *BC*. The crank-shaft, as well as

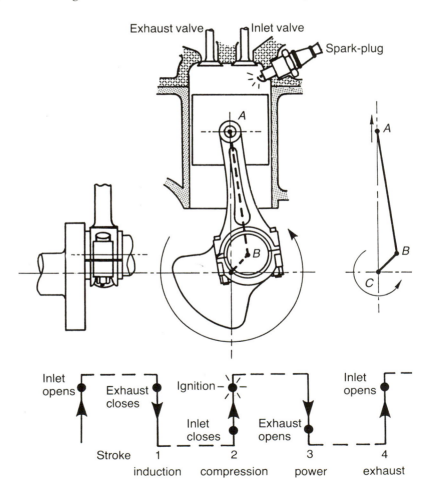

Fig. 1.27 *A four-stroke single-cylinder petrol engine.*

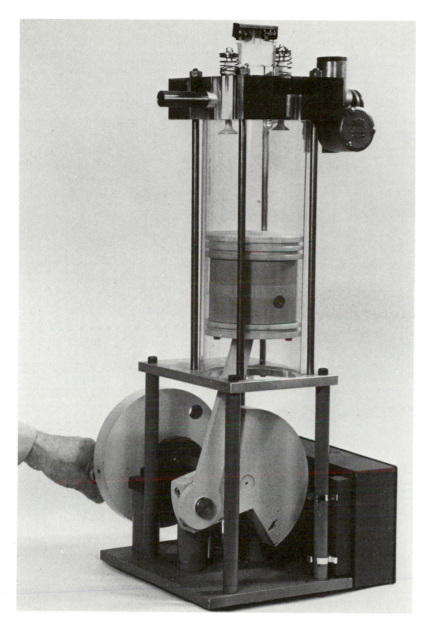

Fig. 1.28 *A musical version of a petrol engine.*

providing the output power, also drives a cam-shaft, usually through a belt, and the cam-shaft in turn moves the valves that control the flow of petrol–air mixture into and out of the cylinder. It is, of course, essential for these valves to be opened and closed at the right instants during each cycle of operation, even when the engine is turning at several thousand revolutions per minute; the various events are shown in the timing diagram.

For the model, Fig. 1.28, everything was moved rather slowly by hand. This involved turning the crank, moving the valves in accordance with the timing diagram shown in Fig. 1.27, and flashing a light bulb inside the cylinder to represent a spark. The sequence started with the inlet valve opening to admit the (imaginary) petrol–

air mixture. As the piston moved down, it drew the mixture into the cylinder, then the inlet valve closed so that the mixture was compressed as the piston rose. The spark ignited the mixture, which expanded and pushed the piston down on the power stroke. Finally, the exhaust valve was opened to release the exhaust. Three volunteers helped, one on each of the valves and the third on ignition. Buzzers emitted loud and distinctive noises when the valves were opened and again when the piston passed its top and bottom positions. The musical performance was very creditable. A cut-away model of a four-cylinder Ford petrol engine demonstrated the real thing.

Linkages made up of bars connected to each other by joints form another important class of mechanisms. They are used extensively in machinery such as textile or handling machines, devices for positioning car bonnets and doors, cranes and robots, and in many other applications where the positioning of parts is best achieved by means of a direct mechanical connection. They are versatile and can be used to move a point along curves or lines, to manoeuvre a body into a chosen position, or to drive a shaft forwards and backwards cyclically.

A popular member of the family is the four-bar linkage, so called because of its one fixed and three moving links. Fig. 1.29(*a*) shows a version capable of following a straight line, as might be required in a crane or a cutting machine. Fig. 1.29(*b*) illustrates its application to the tilting mechanism used in the prototype Advanced Passenger Train, where hydraulic jacks are used to tilt the body of the coach inwards when the train is moving at high speed around curves. Another version is shown in Fig. 1.29(*c*), where a variety of curves can be produced simply by placing the point of the pen at different positions on the extended bar *BC*. Although the geometry may appear simple, the design task – to find the lengths of the links and the position of the joints in order to generate a given curve – is in reality far from straightforward. Direct mathematical solutions are often impossible, and the computer is a great help.

Fig. 1.29 (a) A four-bar linkage. When the smallest link or crank is turned, the point of the arrow on the coupler describes a straight line for part of its path.

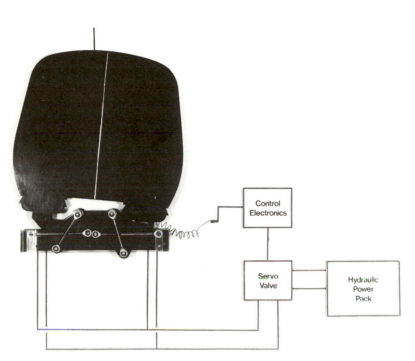

Fig. 1.29 (*b*) *A model of the four-bar linkage used in the tilting system of the prototype Advanced Passenger Train. Courtesy of British Rail.*

Fig. 1.29 (*c*) *A demonstration model fitted with a pen for drawing coupler curves. When the crank is turned by means of the (additional) transparent link, the pen traces out a closed curve: an interesting variety of curves can be drawn by adjusting the position of the pen on the coupler.*

A computer, complete with a suitable program, drew out a four-bar linkage on a visual display unit (Fig. 1.30). By depressing appropriate keys, a volunteer quickly generated a variety of curves marked out by different points on the linkage as it moved, and was also able to vary the proportions of the linkage itself in successive runs. With a program of this kind, a designer can quickly see the effect of adjusting the principal dimensions of the linkage, and so decide whether they are right for his purposes. Computer-aided design, which takes many

Fig. 1.30 *A computerized version of a four-bar linkage.*

Fig. 1.31 *The most advanced mechanism known to man.*

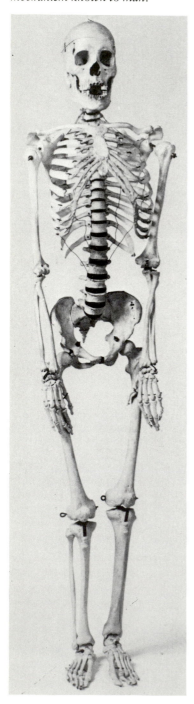

forms, is an increasingly important tool for the engineer.

But the most advanced linkage of all, Fig. 1.31, operates in three dimensions, has more than 200 joints and is capable of precise combined movement under the control of a single computer. A knowledge of how it works is of great importance to us all. We shall look at it more closely in a later chapter.

2

GATHERING MOMENTUM

The great discovery

Almost 2000 years elapsed after Archimedes founded the science of statics in ancient Greece before we began to understand why and how things moved. Many theories were put forward during the intervening centuries to explain the mysteries observed, but the truth of the matter remained unknown until in 1687 Newton related the elusive concepts of force, mass and acceleration in his great central law of motion. It states that a particle accelerates in the direction of the force acting on it with an acceleration directly proportional to the force and inversely proportional to its mass. The units of the quantities involved (force, mass, length and time) are always chosen so as to make the constant of proportionality unity, so that the result can be expressed simply as

$$\mathbf{F} = m\mathbf{a}$$

where $\mathbf{F}$ represents the force, m the mass and $\mathbf{a}$ the acceleration.

A cautionary word about $\mathbf{a}$ is necessary. Because acceleration depends upon displacement, and displacement has to be measured with respect to some frame of reference or another, we have to be careful about the choice of frame. The standard Newtonian frame has its origin at the mass centre of the solar system and has no rotation with respect to the fixed stars; other Newtonian or inertial frames are defined as those that neither accelerate nor rotate with respect to the standard frame. Newton's law holds good in all of them unless, as in particle physics, the velocities in question are close to the speed of light (3×10^8 m/s), when questions of relativity arise. Relativity apart, the restriction to Newtonian frames is important for motions such as navigation in space or the precise motion of gyroscopes relative to the rotating Earth, but for the great majority of engineering applications the errors introduced by assuming the Earth to be fixed are negligible.

One of the difficulties in the way of discovering the laws of motion was the effect of air upon moving objects. When experimentalists measured how fast things fell down, it was not an easy matter to account for aerodynamic forces, or even to recognize that they existed. But if we compare the falling of very different objects *without* aerodynamic complications, we find that they fall with precisely the same acceleration. In each case, the only force is the weight, and, because the weight is proportional to the mass, their accelerations, according to Newton's law, should be the same.

An evacuated glass tube about 1 m long contained some feathers and a coin, Fig. 2.1. When the tube was turned rapidly to an upright

Fig. 2.1 *The glass tube contained a coin and some feathers. In (a), the air had been pumped out of the tube and the coin and the feathers fell under gravity with the same acceleration. In (b), the air admitted to the tube delayed the fall of the feathers.*

a

position and held there, feathers and coin fell from the top to the bottom at the same rate. When air was admitted to the tube and the experiment was repeated, the coin arrived at the bottom well in advance of the feathers, showing the effect of the invisible air.

If, for the *same* body, the force is doubled, the acceleration should be doubled too. Imagine a simple experiment in which an arrow is shot upwards in the air. Let us neglect aerodynamic effects, which for low velocities and a reasonably well-designed arrow will be small. If the string in the bow is pulled back and then released, the initial force on the arrow will have a certain value, and the arrow will have a corresponding initial acceleration. If the string is pulled back twice as far, the initial force on the arrow will be doubled, and its initial acceleration should also be doubled. Things are more complicated during the short interval when the string remains in contact with the arrow; but, provided the weight of the arrow is small compared with

b

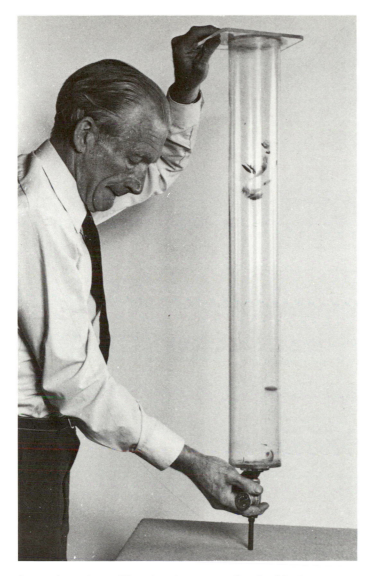

the string force, the string will maintain contact for double the distance in the second case. In other words, everything should be the same in the two experiments except that the unit of length is twice as much in the second as the first (twice the initial acceleration, twice the distance involved in the acceleration phase). It follows that the arrow will leave the bow at twice the speed. The question then is 'How much higher does it go?' Since, in free flight upwards, the *deceleration* in the two cases will be the same, g (neglecting aerodynamics, remember), the faster arrow will take twice as long to get to the top, and, because its average speed will be twice as large, it should go *four* times as high. For a pull three times as large, the arrow should reach a point nine times as high, and so on.

A volunteer agreed to test the strength of this argument. His first shot went up about five feet; the second, with about double the pull, about twenty feet. Both volunteer and lecturer surveyed the descent, secure in the knowledge that the arrow in question had a rubber tip.

If the interval of time during which a force acts is very short and the force is not too large, the change in velocity will be small. As a result, bodies initially at rest will stay almost still, as can be demonstrated by a sudden tug on a tablecloth. Several cups and saucers filled with water were placed on a cloth covering a small table. A sporting volunteer grasped the overhang of the cloth and gave it a sharp horizontal tug, while the audience held its breath. Success! The cloth came away with a slight rattle of china, but not a drop was spilled. The only horizontal force on the saucers, friction, was too small, and its duration too short, for the cups and saucers to gather any appreciable motion. Fig. 2.2 shows a rehearsal in progress.

Fig. 2.2 *Rehearsal in progress. A sharp tug on the tablecloth and no water was spilled.*

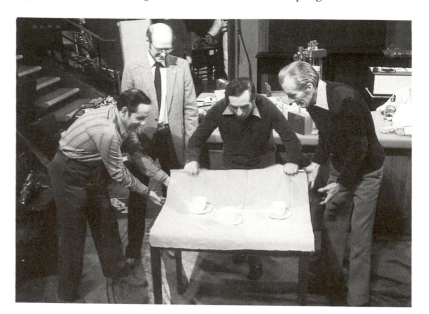

Going round in circles

Circular motion, as text-books show, involves acceleration towards the centre. Even when the particle's speed is constant, the *direction* of its velocity changes as it travels around the circle, and the rate of change of velocity, or the acceleration, is always towards the centre, as shown in Fig. 2.3. A few simple demonstrations can illustrate some interesting features of this kind of motion.

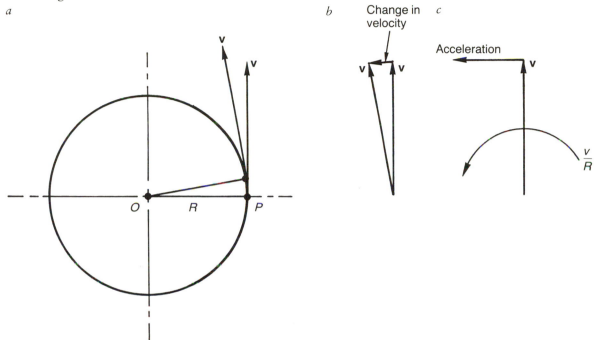

A small rubber ball was whirled around in a horizontal circle at the end of a piece of string, the tension in the string providing the force necessary to maintain the centripetal acceleration. When the string was suddenly released, the central force on the ball dropped to zero. The ball kept travelling in the direction it had at that instant – towards the middle of the audience. It did not, as some might have expected, fly outwards in defiance of Newton's laws.

If the plane of the circle is vertical, gravity may be important, as a toy car looping a loop inside a plastic track can demonstrate, Fig. 2.4(*a*). At the top, the central force is made up of its weight, acting downwards, plus the track reaction, also acting downwards. If the speed of the car is high, the centripetal acceleration, which depends on the square of the speed, is also high and the track reaction must reinforce the weight to provide sufficient central force. But, at a certain low speed, the weight of itself can provide enough force, and at lower speeds yet the track reaction would have to be *upwards* to achieve the correct force. This is impossible, and the car falls out of circuit.

A more interesting version of the same idea was demonstrated by whirling a bucket full of water around a vertical circle, Fig. 2.4(*b*).

Fig. 2.3 *Motion of a point around a circle at constant speed. In a short interval of time, the velocity of the point P changes its direction, and the change determines the average acceleration. To find the* instantaneous *acceleration, (c) shows the velocity vector turning at rate (v/R). The rate of change of the velocity, or the instantaneous acceleration, is the velocity of the tip of the vector: its direction is perpendicular to* v *(towards the centre of the circle) and its magnitude is (v²/R).*

Fig. 2.4 *Motion around a vertical circle. In (a), the model car was travelling too slowly, and fell out of circuit. In (b), the water did not always stay in the bucket.*

a

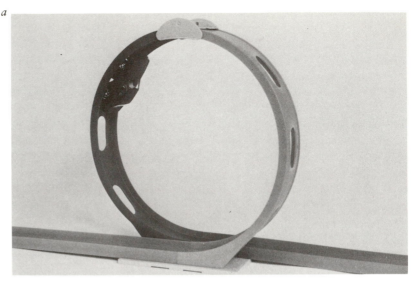

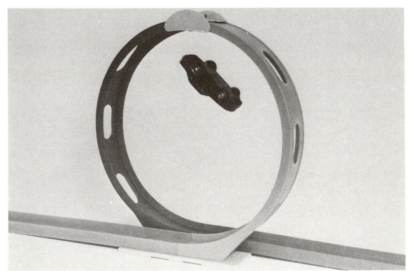

b

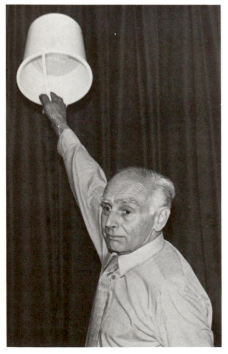

When the demonstrator swung the bucket, he was asked to swing it as *slowly* as possible without causing a deluge. It was surprising to find that the slowest rate was so slow, but in the interests of science he did it even more slowly and was rewarded with a brisk shower.

Consider another kind of circular motion in which a bob whirls round a vertical axis in an arrangement like that of Fig. 2.5(*a*). The resultant force on the bob, made up of tension in the inclined string and its own weight, acts towards the centre of the circular path. For a faster turn and a correspondingly higher centripetal acceleration, a larger central force is required to maintain steady motion and both the string tension and the inclination of the string will increase. In other words, the inclination of the string θ depends on the angular velocity at which the body whirls around the vertical.

Fig. 2.5(*b*) shows a model of how this idea is incorporated in one of the oldest devices for controlling the speed of rotating machinery, the centrifugal governor. It was used by James Watt in his steam engines and still has many useful applications today. Instead of the string and a single bob, a linkage supports at least two fly-balls. The entire assembly is driven around a fixed vertical via a direct mechanical connection from the engine, so that its speed of rotation is proportional to engine speed. If, for some reason, the engine speed exceeds the required value, the fly-balls move outwards; as they do so the linkage compresses, sliding a sleeve upwards along the vertical driving shaft. This in turn cuts down the fuel supply to the engine, and the speed falls to its previous value. The opposite happens when the engine speed becomes too low.

a

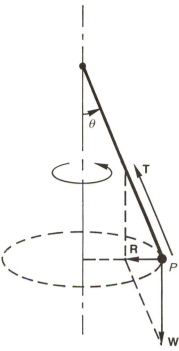

b

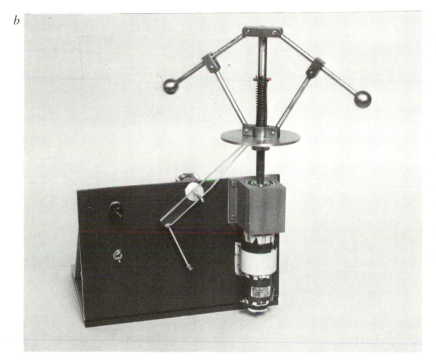

Fig. 2.5 (a) *When the particle P at the end of a light string describes a horizontal circle at constant speed, the resultant* **R** *of its weight* **W** *and the string tension* **T** *points towards the centre of the circle.* (b) *A model of a centrifugal speed governor.*

Angular motion, Fig. 2.6 is based on the same principles as linear motion. Instead of the driving *force*, we have to consider the driving *torque T*, which is the product of the magnitude of the force and the perpendicular distance from its line of action to the axis of rotation. The angular acceleration α is proportional to this torque. And, instead of the mass of the body, we have the *moment of inertia* of the body about the axis, I, a quantity that depends on the *distribution* of mass. A body with a lot of mass further out has a moment of inertia larger than that of a body with its mass concentrated near the axis, even if the total mass is the same: for the same torque, its angular acceleration will therefore be less. The relationship can be expressed concisely as

$$T = I\alpha$$

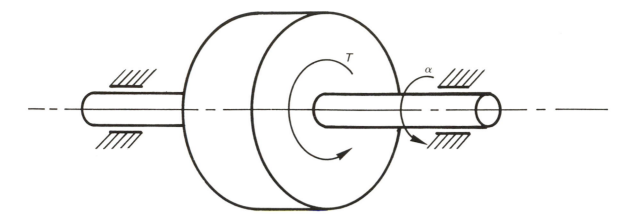

Fig. 2.6 *Angular motion. The torque* T *causes an angular acceleration* α.

To take matters further, suppose that the body in question has no fixed axis of rotation at all, but is free to move in a plane, like a flat bar sliding on ice. Then the most important point in the body is its centre of mass, often called its centre of gravity and always given the letter *G*. If the body is perfectly free to move, a force applied on one side of *G* will make it turn in one direction and a force applied on the other side will make it turn the other way.

The centre of mass of a wooden pointer about 1 m long was marked with coloured tape, Fig. 2.7. Initially the pointer was held lightly in a vertical position on a smooth table, so that its weight was closely balanced by the vertical reaction of the table. Friction at the table was small, so that during its initial motion the conditions approximated to free motion. When the bar was given a sharp horizontal blow to the right of the picture at a point above *G*, the bar turned *clockwise* as *G* moved to the right. When the experiment was repeated with a similar blow below *G*, it turned *counter-clockwise*, G still moving to the right. Finally, when the blow was delivered exactly *at G*, it remained upright – until it left the table altogether and was neatly gathered by a volunteer waiting in the slips.

Fig. 2.7 *A horizontal blow delivered below* G *causes a rotation in a counter-clockwise direction, viewed from the front: above* G, *and the pointer rotates clockwise. In each case,* G *moves in the direction of the blow. When struck* at G, *the pointer began to move without any rotation.*

Mighty thrusts

Perhaps the most dramatic examples of man-made motions are those of space rockets, propelled by the thrust of gas expelled from the rocket engines, as shown in Fig. 2.8(*a*). The basic principles of their dynamics follow directly from Newton's laws but, because the rocket structure and its exhaust are clearly not a single rigid body, it is impossible to think in terms of a single acceleration. Instead, we can make use of the second law expressed in terms of a quantity called *linear momentum*, which is defined as the product of mass and velocity summed over all the masses in the rocket and the exhausting gases. The alternative form of the law asserts that the external force acting on the entire system is precisely equal to the rate of change of its total linear momentum, a version that is entirely equivalent to the usual statement when the system is a rigid body. If,

Fig. 2.8 *Rocket propulsion. (a) The first launch of the European space rocket Ariane. (b) A smaller version in the lecture theatre, driven by compressed air and water.*

a

b

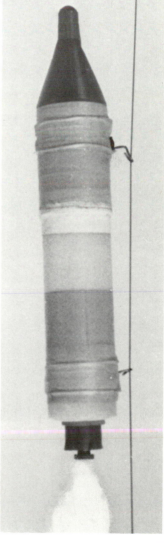

to simplify matters, all external forces on the rocket are disregarded, it follows that its total linear momentum remains constant, i.e. the *backward* linear momentum of the exhausting gases must be accompanied by a *forward* linear momentum of the rocket structure. Although the engineering of a practical rocket is complicated, the principles of rocket thrust can be demonstrated without difficulty.

First, several long balloons were released from the audience. As they deflated, they obligingly performed a variety of erratic and noisy manoeuvres, in strict accordance with the principles of linear momentum. Next, two small rockets containing a mixture of pressurized air and water, one of which is shown in Fig. 2.8(*b*), shot up towards the ceiling on guide-wires, leaving sprays of water behind. As might have been expected, the rocket containing the larger mass of water, and thus the greater capacity for thrust, rose to a more impressive height than its companion.

Finally, a special chair fitted with air-pads on its four feet was brought into play. The air-pads were supplied with pressurized air that supported the chair fractionally above the large glass plate, rather like a hovercraft. Fig. 2.9(*a*) shows a rehearsal in progress. Friction between the chair and the glass was very small, so that when the demonstrator sat on the chair, equipped with two small fire extinguishers, he found he could glide easily in any direction. A short burst from the extinguishers, directed forwards, drove him backwards satisfactorily, and a similar burst backwards drove him forwards. When a volunteer took his place with a single extinguisher held at arm's length, the backward motion was accompanied by a rotation going in the opposite direction from the jet. Such a technique is used to control the angular position of satellites, although in that application the pulses of gas, emitted from tiny jets like those of Fig. 2.9(*b*) and strategically placed around the structure, must be carefully metered to achieve accurate results.

a

Fig. 2.9 *Jet manoeuvres. (a) A hover chair equipped with fire extinguishers. Courtesy of the National Engineering Laboratory. (b) A tiny jet nozzle for positional control of satellites. Courtesy of British Aerospace.*

b

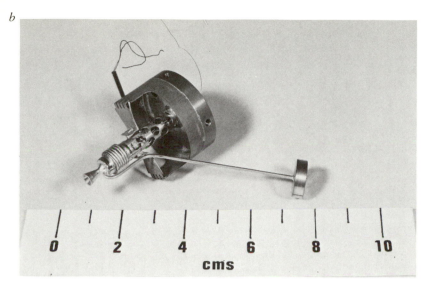

We shall be looking at jet-engines later, but for the time being consider them simply as devices that produce thrust. The most versatile application of thrust control is found in that famous machine, the Harrier, in which the direction of the thrust is controlled by the pilot by means of rotatable nozzles, two on each side of the fuselage, as shown in Fig. 2.10(*a*). By varying the direction of the nozzles from straight back for forward propulsion to straight down for lift, the pilot can perform the most extraordinary manoeuvres ever achieved by a fixed wing aircraft, including transition from jet-lift in hover to aerodynamic wing-lift in forward flight. Because the weight of the aircraft is less than the thrust available from the Pegasus engine, there is adequate thrust for stationary hover; but, if the aircraft is not to topple in pitch, the two forces, weight and thrust, must be accurately in line, Fig. 2.10(*b*). This formed one of the most important

a

Fig. 2.10 *The Harrier. (a) Vertical take-off with the main nozzles directed downwards for lift.*

Fig. 2.10(b) *For stability, the resultant thrust must act through the centre of gravity* G.
(c) *Additional control is provided by small nozzles at the wing-tips and at the nose and tail. Courtesy of British Aerospace.*

b

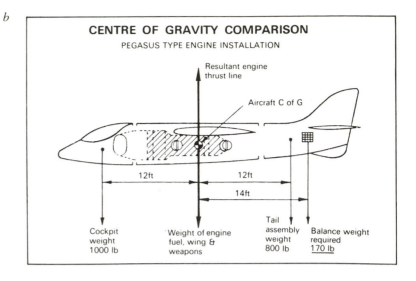

c

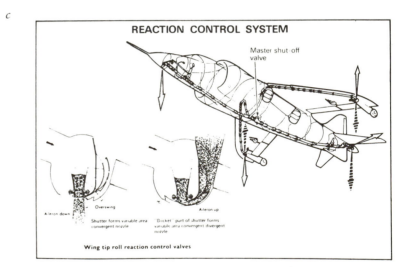

considerations of the design – the resultant thrust had to act through the centre of mass of the aircraft. In addition, small nozzles fed from the engine are placed at the wing-tips and at the nose and the tail to enhance control, Fig. 2.10(*c*).

Sudden shocks

Ideas of impact are closely connected with momentum. In impacts or collisions, the forces involved are very large and the action is completed in a very short interval of time. This often makes it difficult to establish the details. Instead, the states of motion just before and just after the impact may be compared with a view to predicting the course of events.

Newton's cradle, Fig. 2.11, is a good example. Seven identical steel ball-bearings, suspended from a wooden frame by light strings, could

Fig. 2.11 *Newton's cradle. A variety of interesting motions can be generated by releasing different numbers of balls from either side.*

swing in a vertical plane like simple pendulums. At rest, they hung straight down, each just touching its neighbours in a straight line. When one of the end balls was raised at an angle and released, it swung downwards until it struck its stationary neighbour with a horizontal impact. Almost immediately, the ball at the far end of the line swung outwards, leaving the other five, together with the striker, stationary. After the moving ball completed its swing, first outwards then back, it struck its neighbour at virtually the same speed as in the original impact, and the cycle of events was repeated many times. When *two* balls were raised initially, leaving five stationary, *two* at the other end joined in the new pattern of cyclic motion; when three were started off, three bounced off the other end . . . and so on. The audience was invited to speculate what would happen when four balls were raised, so that only three were left to receive the collective impact. Several observers got it right. After the first impact, the leading ball of the quartet joined forces with the stationary trio for the first outward swing, led the reverse swing, and then rejoined its original group after the second impact.

Many interesting patterns of motion can be achieved with this simple arrangement. Two principles of mechanics are involved. First, because no external force acts horizontally on the balls during impact, the horizontal linear momentum of the entire set must be the same just after the impact as it was just before. Thus if a single ball is released and its linear momentum just before impact is (mv), the total linear momentum of the set after impact must also be (mv). This will be the case if the ball at the far end starts off with the same velocity as the striker and all other balls including the striker remain stationary. But an equally satisfactory explanation would apply if *two* balls at the far end started off after impact, each with half the original velocity –

or even three with a third of the velocity. Clearly something else is involved. That something is kinetic energy. Because the steel balls are almost perfectly elastic, very little energy is lost in impact, so that the total kinetic energy ($\frac{1}{2}mv^2$) is practically the same after impact as before. It is a nice mathematical exercise to show that the only way both linear momentum and kinetic energy can be conserved is in the way things actually happened.

Unfortunately, impacts involving people occur all too often on our roads, and the design problem of protecting human life and limb inside steel boxes travelling at high speeds continues to exercise the skills of many engineers. Their approaches take several forms. They include improvements in the structural design of the vehicle to confine as much damage as possible to the structure, and restraint of the passengers by means of seat belts. Many complicated factors are involved. The belts must be positioned and anchored around the body in the best possible way, they must be sufficiently strong, they must provide a degree of restraint that is neither too stiff nor too soft, and they must be usable by passengers of many different shapes and sizes. Altogether this is a tall order in which both experiment and analysis are playing important parts. Dramatic reductions in injury have already been achieved. Fig. 2.12(*a*) shows an engineer's definition of the many geometrical factors involved. In Fig. 2.12(*b*), a heavily instrumented car is shown just after a head-on impact with a concrete barrier; the importance of such testing for purposes of safe design can hardly be exaggerated.

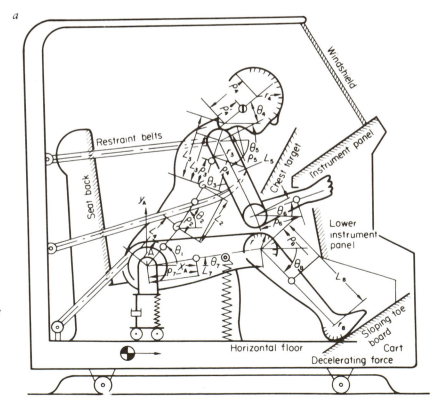

Fig. 2.12 *Impacts in cars. (a) The geometry of the driver. Courtesy of Bioengineering Unit, University of Strathclyde. (b) A test in progress. Courtesy of Motor Industry Research Assocition.*

b

An uncooked egg, packed around with paper in a stiff cardboard box, was thrown vertically upwards several times to heights of about 3 m and allowed to fall to the floor. When unpacked, the egg was found to be intact, protected by its paper nest. Without such restraint, it would assuredly have been smashed. A demonstration of a similar kind was provided by a robust wooden rod supported horizontally on two fragile wine glasses placed on stools about 2 m apart, Fig. 2.13. When the demonstrator, armed with a sword, delivered a mighty

Fig. 2.13 *A mighty blow that broke the wooden rod but not the wine glasses.*

blow downwards at the middle of the rod, it broke in two; the wine glasses, although left tottering slightly, were unharmed. The trick was to support the shank on the glasses by means of ordinary pins sticking outwards at the ends of the shank. The shock of the blow, large enough to snap the wood, was thereby prevented from reaching the glasses, illustrating the advantage of a soft connection.

Gyroscopes and curious motions in the third dimension

There is a close analogy between the laws governing linear motion on the one hand and angular motion on the other. In the same way that *force* acting on a system can be equated to the rate of change of *linear* momentum, *torque* can be equated to the rate of change of *angular* momentum. In considerations of angular motion, the axis about which rotation occurs must first be established – an obvious enough matter if the body is turning in fixed bearings, but not so obvious for something sliding and turning on a sheet of ice, or for an aircraft manoeuvring in three dimensions. Although the mathematics of three-dimensional motion is difficult, even for a single body, the approach can be made more manageable by a judicious choice of axes.

At each and every point in a body, there are three perpendicular axes, called the principal axes of inertia, which possess a remarkable property: the moment of inertia of the body about one of them is the largest and about another it is the smallest, in comparison with all other axes through the same point. With reference to these axes, Fig. 2.14, the three perpendicular components of angular momentum of the body can be expressed simply as the product of the appropriate moment of inertia and the corresponding component of angular velocity, a result that is completely untrue for any other set of axes. Moreover, dynamical theory shows that if the point in question is fixed (such as the tip of a spinning top) the torque acting on the body is precisely equal to the rate of change of total angular momentum,

Fig. 2.14 *Principal axes of inertia and the components of angular momentum.*

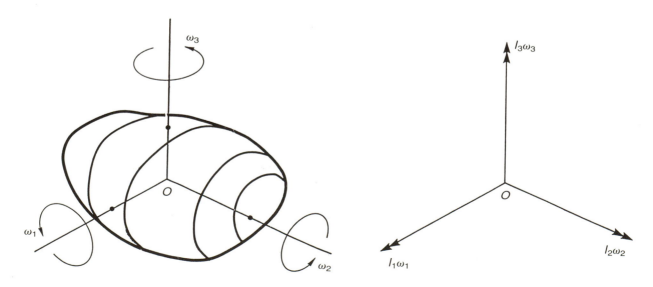

made up of the separate contributions of the three components. Despite the complications of the third dimension, this approach greatly simplifies the analysis of the motion. Fortunately, the same approach holds good even if *no* point of the body is fixed, as in the manoeuvring aircraft, provided the point of reference is taken to be the centre of mass. Once again, G is marked out as the most important point in the business.

These ideas apply to one of the most remarkable manifestations of angular momentum, the gyroscope. It is not only a fascinating toy, but it also forms the basis of a wide range of practical instruments. A gyroscope is simply a spinning body, usually having the form of a circular wheel or rotor, held in such a way that the axis about which it spins is free to turn in space. This arrangement can result in some very peculiar motions. Paradoxically, however, its most useful feature is the marked reluctance of its axis of spin to turn at all. Fig. 2.15 explains this strange behaviour. There a disc is shown miraculously suspended in space at its centre and spinning about its axis of symmetry Oz. Its angular momentum can be represented by a vector pointing along that axis. Suppose an external torque is applied to it about a diametral axis Ox. The torque can be represented by another vector, this time along Ox. Remembering that the torque vector must be equal to the *rate of change* of the angular momentum vector, we conclude that the latter must turn such that the velocity of its tip is in the same direction as the torque vector. It follows that the disc must turn about the third, perpendicular, axis Oy – contrary to normal experience of the way things turn under torques. Such a motion is called precession, the second most important characteristic of a gyroscope. Moreover, the larger the angular momentum vector, the slower is the precession: in other words, the higher the rate of spin,

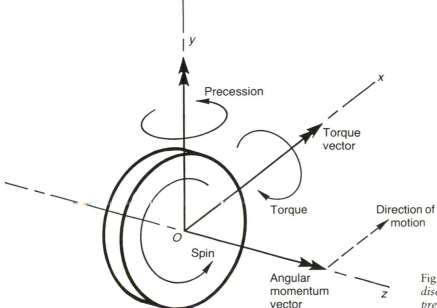

Fig. 2.15 *Precession of a spinning disc. A torque about* Ox *causes precession about* Oy.

the more stable the spinner.

The most familiar example of such stability is the Earth, which spins rather slowly compared with machinery but which, because of its large polar moment of inertia, possesses a vast angular momentum. As the Earth progresses on its annual transit around the Sun, its axis maintains a direction almost fixed in space, as illustrated in Fig. 2.16.

The advantage of spin in stabilizing the direction of a body also finds application in bullets or shells fired from a rifled gun-barrel, where the spiral rifling or grooving of the barrel gives the projectile a rapid spin as it leaves the muzzle. Some satellites are stabilized by spin. In quite another field, rugby players have been known to spin the ball to improve the accuracy of their aim. Curiously, spin is not *always* stabilizing; if applied wrongly, it will have the opposite effect. The explanation of this peculiar result depends on the three principal

Fig. 2.16 *The spin of an exaggerated Earth keeps it pointing in a fixed direction.*

axes of inertia mentioned previously. Dynamical theory shows that if spin is applied about the axis of either the largest or the smallest moment of inertia, the motion will continue in a steady fashion provided only that the centre of mass lies on the axis in question. But if spin is applied about the intermediate axis, the result will be an unsatisfactory and progressive wobble.

A thick disc of heavy wood was hung on a long thread attached to the disc at its centre of mass, Fig. 2.17. It was held at rest in a horizontal position and then given a brisk spin about the line of the thread. As it wound up the thread, it continued to spin stably and remained perfectly stable even after the spin reversed when the thread unwound. Two identical steel bars were then fitted into vertical holes drilled in the disc, the centre of each bar being placed accurately at the mid-thickness of the disc. The holes were diametrally opposed at equal distances from the centre, so that the centre of mass of the assembly remained at the centre of the disc. When the disc was again spun about the line of the thread, the ensuing motion was distinctly unstable and the disc wobbled increasingly until it tangled with the thread. This strange behaviour depended on the changes in the moments of inertia introduced by the rods: when the rods were fitted,

a *b*

Fig. 2.17 *Spin does not always stabilize. In (a), the axis of spin was that of the largest moment of inertia and the motion was stable. In (b), the same axis was that of the intermediate moment of inertia, and the spinner wobbled unstably.*

the line of the thread became the axis of intermediate moment of inertia, whereas previously it had been the axis of maximum value.

Spinning gyroscopes are used in many different forms. Often the rotor is held in two frames, or gimbals, so that its axis can adopt any direction in space, as in Fig. 2.18(*a*). Alternatively, if the base is attached to something that is turning slowly, such as an aircraft that is banking or pitching, the rotor's axis will keep pointing in a fixed direction, and an observer can deduce how much the aircraft has turned from the instrument itself. This arrangement is widely used in gyro-verticals for short-term manoeuvres, where the accelerations involved would render an ordinary pendulum useless.

In another form shown in Fig. 2.18(*b*), the rotor is mounted in a single gimbal, which enables the gyroscope to measure a rate of turn. As the base slowly rotates about a vertical axis, the gyroscopic action of the rotor causes the gimbal to adopt a definite angle against the action of a spring, and the angle provides a measure of the rate of turn. Instruments of this kind have been developed to high degrees of performance for use in the navigation and control systems of aircraft and ships, where they are capable of measuring rates of turn a tiny fraction of the Earth's rate.

The rotor of a double-gimbal gyroscope was driven up to speed by

Fig. 2.18 Gyroscopes. (a) A so-called two-axis gyroscope in which the spinning rotor is held in two frames or gimbals, each capable of turning with respect to the other and to the base. (b) A single-axis gyroscope in which the rotor is held in a single gimbal. Courtesy of Academic Press.

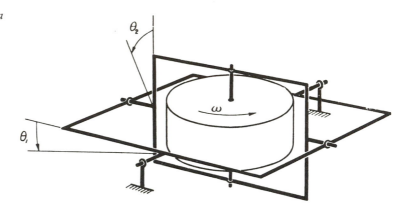

a friction disc attached to a small electric motor, as in Fig. 2.19. When the speed reached several hundreds of revolutions per minute, the outer gimbal was given a sharp blow with a hammer. Contrary to normal experience, the gimbal did not revolve in its bearings, as it assuredly would have done if the rotor had had no spin, but instead it held its position – apart from a slight shudder – with impressive stability. Repeated blows confirmed the matter. When a steady push was applied, so as to create a torque about the gimbal axis, the gimbal still refused to move, but now the *inner* gimbal turned on its perpendicular axis – corresponding to the precessional motion discussed above. When the push was applied in the opposite direction,

Fig. 2.19 *With the rotor spinning well, the gimbal resisted a sharp blow with the hammer although it was free to rotate.*

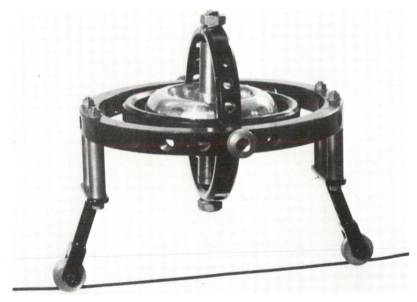

Fig. 2.20 *Peculiar motions.* (a) *The spinning rotor preserved excellent stability for a tight-rope excursion.*

the inner gimbal obligingly returned to its central position. In theoretical terms, the torque applied was equal in magnitude and direction to the rate of change of angular momentum.

Another convincing demonstration of stability was provided by a small gyroscope consisting of a rotor spinning in a double gimbal ring, Fig. 2.20(*a*). Two small wheels, arranged as in a bicycle, could support the weight of the assembly, and when the wheels were placed on an inclined wire the model rolled down the wire in a perfectly upright position until it reached the bottom.

Finally, a large top in full spin was placed on a fixed support at an

Fig. 2.20 (b) Precession squared: the tops precessed slowly under the action of torques exerted about their tips by their weights.

Fig. 2.21 Count Schilovsky's two-wheeled car (1914) stabilized by an internal vertical-axis rotor. It made several demonstration runs including one from London to Birmingham. Courtesy of Hutchinson, 1954, from The Gyroscope Applied *by K.I.T. Richardson.*

inclination to the vertical. A non-spinning body would immediately have fallen down, but the spinning top precessed in a dignified manner around the vertical at a fixed inclination. When a second, smaller top was placed precariously on the first, the pair performed an interesting *pas de deux* in a perfectly stable manner, as shown in Fig. 2.20(*b*).

Although gyroscopes are now used in practice to detect rotations rather than to apply torques, inventors in the early years of the 20th century were busily engaged in using large rotors to stabilize bodies directly. Their inventions included ship stabilizers, which employed a massive gyroscope to limit rolling action, and two-wheeled vehicles – one wheel in front, the other behind – to run on a single rail or even on the road. Fig. 2.21 shows a gyro-stabilized two-wheeler. Such devices were technically feasible, but the dreams of their inventors were never realized: the world at large preferred at least four wheels. Various designs of ship stabilizer incorporating massive spinning rotors were successfully developed and saw good sevice at sea, but it was not long before more competitive arrangements, incorporating underwater fins activated by a small sensing gyroscope, rendered them obsolete.

Amongst the most advanced types of gyroscopes so far developed are the tiny instruments used for purposes of navigation and control in aircraft, space vehicles and ships. They are used to detect minute rotations of a directionally stabilized platform carried inside the vehicle in such a way that the platform can maintain fixed directions in space no matter how the vehicle turns. Although the mechanical arrangements and the control systems are complicated, the engineering is of such a high order that departures of the platform from the chosen directions can be reduced to a tiny fraction of a degree over many hours of active operation. Such reliable

a

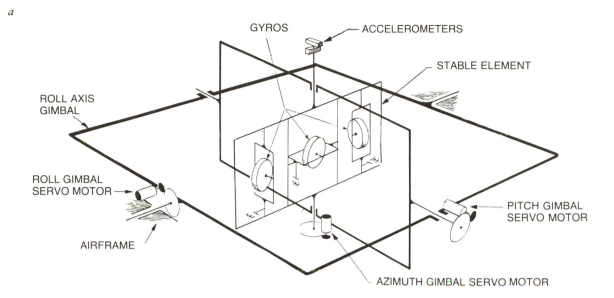

GYROS — ACCELEROMETERS

STABLE ELEMENT

ROLL AXIS GIMBAL

ROLL GIMBAL SERVO MOTOR

PITCH GIMBAL SERVO MOTOR

AIRFRAME

AZIMUTH GIMBAL SERVO MOTOR

performance can be achieved only with scrupulous attention to the details of manufacturing. Special clean rooms are employed in which the temperature is maintained constant, and all air filtered to eliminate the tiniest specks of dust or lint.

The stable platform shown schematically in Fig. 2.22(*a*) forms the heart of the navigating system. It consists of the central stabilized platform S, mounted in two gimbal frames which allow it to remain directionally fixed in space as the carrier turns. The platform carries three sensitive gyroscopes upon whose performance the assembly critically depends. Any small departures of the platform from the required directions are immediately detected by the gyroscopes, which thereupon issue signals to miniature electric motors M_1, M_2 and M_3, which in turn hold the platform fixed. To complete the assembly, another set of instruments called accelerometers are mounted on the stable platform. Their function is to measure the instantaneous acceleration of the platform as it is carried by its transporting vehicle over the surface of the Earth or in space. Armed with this information expressed along the known directions of the platform, a computer can rapidly determine all the navigational information that is required, including the velocity, the position and the course of the vehicle. Moreover, it can do so without any external references whatever: all is self-contained. This remarkable engineering application of Newton's laws is now standard equipment in all large aircraft, usually in triplicate. The sectioned view of an actual assembly, illustrated in Fig. 2.22(*b*), shows the degree of compactness achieved in practice.

Not all instruments called gyroscopes consist of a spinning rotor. Much ingenuity has been applied to develop other ways of detecting and measuring small angular motions, including vibrating mechanical elements, rotating fluid masses and magnetic fields. A recent arrival on the scene is the ring-laser gyroscope. Essentially, it consists of a laser beam – a thin ray of concentrated light having a single frequency –

Fig. 2.22 (a) *A schematic drawing of a directionally stabilized platform. Any small departures of the central frame from fixed directions are detected by the gyroscopes, which activate electric motors so as to restore the status quo.*

Fig. 2.22 (b) *Components of a real inertial navigator. From the left: a gyroscope, the stable platform, the complete assembly. Courtesy of Ferranti.*

split into two rays that are sent around a closed path in opposite directions by means of reflecting mirrors and then recombined, as illustrated in the triangular tube on the right of Fig. 2.23. If the platform on which the laser ring is mounted is rotated about an axis perpendicular to the plane of the ring, the frequencies of the two rays change fractionally, increasing slightly on the one hand and decreasing on the other. The difference can be measured by reference to the patterns of light and dark bands, or interference fringes, caused by the difference in their frequencies. Fortunately, the physical quantities involved admit a practical result. For helium–neon lasers having a frequency of 4.74×10^{14} Hz, the difference in frequency associated with an equilateral triangular ring of length 43 cm is 9.51 Hz for a rate of turn of 15°/h (Earth's rate). In principle, measuring this frequency is an easy matter, so that the measurement of very slow rates of turn is well within reach. As usual, however, although the physical principles were well established in the early stages of development, their realization in practice posed difficult problems of engineering.

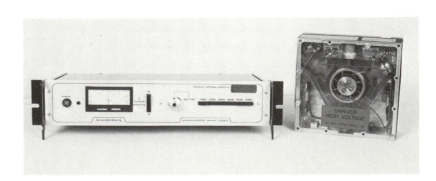

Fig. 2.23 *A ring-laser gyroscope on the right, instrumentation on the left. Courtesy of Ferranti.*

3

VIBRATION

Vibrations great and small

Vibration in machines is usually harmful and often dangerous. The damage it may cause is all too familiar – clashing of parts, noise, wear, deformation and possibly destruction. As many machines are used to carry people around by land, sea and air, and others are made on the heroic scale, it is not surprising to find that extreme care is taken in their design and testing. Lives may be at stake.

During the past 50 years, the speed of machines has increased enormously, and so has their potential for vibration. This has made the task of suppression very challenging. Earlier machines were often designed on the principles of statics, but as performance improved it became essential to take dynamical factors into account. That, as often as not, meant vibration. Even now, unexpected things can happen, and it is not always easy to track down the culprit. Skilled detective work may be required, such as in the following true-life story of a large and powerful machine.

The victim was a large steam turbine similar in size to that shown in Fig. 1.1(*c*). Because the total weight of the rotating parts of these machines is several hundred tonnes, the problem of ensuring smooth running at 3000 rev/min is obviously of some importance. Engineers normally check performance by monitoring vibration via instruments installed at the bearings, and the evidence in this case was peculiar. After a long period of smooth running, the machine was shut down for routine inspection and then started up again. The record of the ensuing events is shown in Fig 3.1.

At first all went well. There was a small vibration, but nothing out of the way. The trouble started after about an hour and a half, when

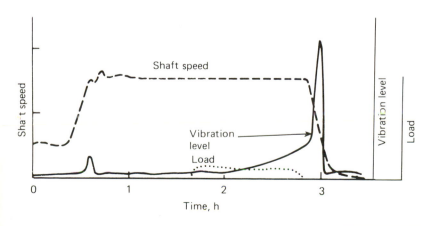

Fig. 3.1 *The record of a mysterious vibration of a large steam turbine shaft. Courtesy of the Central Electricity Generating Board.*

Fig. 3.2 *Useful vibrations.*
(a) A vibratory bowl conveyor;
(b) a pneumatic hammer.

a

b

the electrical load was connected. Vibration increased and went on increasing. After another half hour, the vibration began to grow very rapidly, even after the load was disconnected. It soon became intolerable and the machine was shut down. The same pattern of events was repeated on the second start-up. In view of the long periods involved, the investigators realized something unusual was afoot, but the culprit was by no means obvious. But they had observed that the frequency of the vibration was the same as the speed of the machine, which suggested unbalance of some kind, and it was concluded that the massive steel shaft was slowly bending as the load was increased. The problem was to identify the cause.

Suspicion narrowed down to thermal effects, and the final search began. It converged on a small hole that had been bored along the centre of the shaft to remove impurities in manufacture. When its end-cap was removed, the culprit was found inside – about two pints of lubricating oil. Somehow the oil had found its way into the hole and had been sealed inside. When the machine was cold, the oil remained harmlessly in liquid form, but as the temperature went up it vaporized and extracted heat from one side of the hole, thereby bending the shaft. When the oil was removed and a new start was made, all was well. A meticulous piece of detective work had paid off handsomely.

Fig. 3.2 shows two machines in which vibration serves a useful purpose. The bowl conveyor is made to vibrate in a combined vertical and torsional mode by means of powerful electromagnets mounted inside its base. A narrow spiral track running up its inside surface provides a path along which small objects such as nuts are driven in an orderly way as a result of the vibration; a heap of nuts poured into the bottom of the bowl will be found to travel upwards to an exit at the top of the spiral. When they drop out, one by one, they are well placed for the next stage of production.

The familiar pneumatic hammer is another interesting if noisy example of useful vibration. Inside its body, the compressed air supply is interrupted many times per minute and a succession of pulses is delivered to the steel tool or bit, which is then used to break up hard materials such as concrete or road surfaces.

Despite their apparently straightforward action, good design of these machines is not a simple matter. In the case of the bowl conveyor, for example, decisions have to be made about the shape and size of the spiral, the friction at its surfaces and the kind of vibration to be applied.

To describe vibration, certain terms are essential. We have already met the *frequency*, the number of cycles executed per second. The *period* is the time taken to complete one cycle; if the frequency is 2 Hz (one hertz, or 1 Hz, means one cycle per second), the period is 0.5 s. The *amplitude* is the maximum travel from the central position, and the *phase*, usually expressed as an angle, relates to the timing of the cycle.

To illustrate these terms, consider the projection Q on a diameter of a point P travelling at constant speed around a circle, Fig. 3.3. Point Q executes a simple vibration from end to end of the diameter as P goes round the circle. If the rotational speed of OP is 10 rev/s, the frequency of Q's vibration is 10 Hz. The amplitude of vibration equals the radius of the circle. If a second point R travels around another circle of the same centre but with OR lagging OP by a fixed

Fig. 3.3 *Simple harmonic motion. (a) As OP turns at a constant rate, the projection Q of P on the vertical diameter vibrates with simple harmonic motion: if OR lags behind OP by a fixed angle α, the vibration of S has the same frequency as that of Q but a different amplitude and a phase lag α. (b) The two curves show how the displacement of Q and S vary with time.*

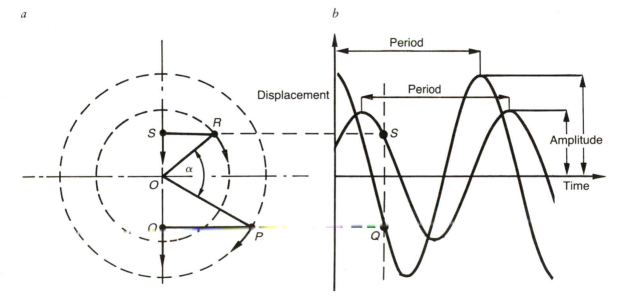

a b

angle α, the vibration of the projection S will clearly have a different amplitude (the radius of the second circle) and it is said to have a phase-lag α with respect to Q: for example, if R is directly opposite P, so that Q and S move in opposition, the phase-lag is $180°$ or π radians. The ways the displacement of Q and S vary with time is called simple harmonic motion and is shown in Fig. 3.3(*b*).

Not all vibrations are so simple. The torque exerted on the crankshaft of normal four-stroke petrol engines by the gas forces inside a single cylinder varies somewhat as shown in Fig. 3.4. A high peaky torque in the power stroke is followed by low values during the exhaust and induction strokes, and the torque actually reverses during compression. This is a difficult curve to express mathematically, but

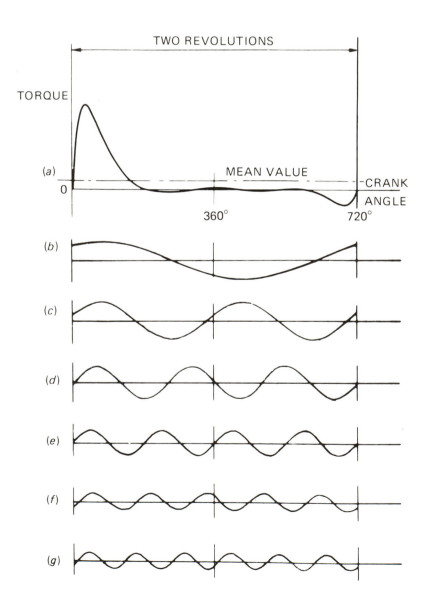

Fig. 3.4 *Variation with crank angle of the gas torque from a cylinder of a four-stroke petrol engine. For purposes of analysis, the curve (a) representing the actual torque can be expressed as the sum of a series of simple harmonic curves (b), (c), (d) . . ., which is much more convenient mathematically.*

fortunately help is at hand. The famous mathematician Fourier showed how *any* periodic curve could be represented by breaking it down into simple harmonic components, each with its own frequency, amplitude and phase. The result for the gas-torque curve is shown in Fig. 3.4: strange as it may seem, the addition of all the curves ($b + c + d + \ldots$) – only the first few are shown – results in the original curve (a). By such means, analysis is greatly simplified. Even so, when the designer is considering lay-outs of several cylinders to achieve smooth torque transmission, he still has plenty to think about.

Free vibration

A simple example of a vibrating system, used more often than any other by engineers, consists of a single body suspended on a light spring, Fig. 3.5. If the body is pulled down from its static position and then released, it will vibrate up and down at a definite frequency. This kind of vibration is called *free* vibration, and its frequency, called the

Fig. 3.5 A simple and useful idealization of a vibratory system consists of a single rigid body vibrating in one direction only, suspended on a spring of zero mass. No known system answers this description exactly, but many are close approximations.

natural frequency, depends only on the mass of the body and the stiffness of the spring. It can easily be calculated. Curiously enough, the natural frequency does *not* depend on the amplitude of vibration; if we double the initial downwards deflection of the body, the frequency of the ensuing vibration will remain unchanged – in other words, the natural frequency is a property of the system, not of the motion. This little model is of great importance, and it can often represent complicated structures in a very convenient way.

Inevitably, if the spring is stretched too far, complications occur. The restoring force of the spring fails to remain proportional to its extension, and the frequency of free vibration is found to depend on amplitude. The theory of such non-linear vibration is much more difficult, but fortunately many practical problems involve small amplitudes only and linear theory can be applied successfully.

There is another complication, too, although in practice it is usually welcome. If a free vibration is left to continue untouched, it will gradually die down and finally stop as various frictional forces of the air, of the support and of the spring dissipate small amounts of energy in each cycle. Typical decay curves are shown in Fig. 3.6. Such frictional or damping forces play an important role in limiting vibration in many machines: shock absorbers in car suspensions are a familiar example.

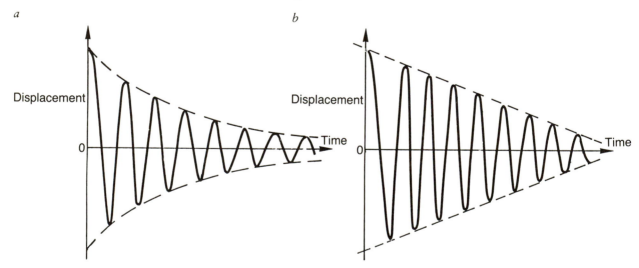

Fig. 3.6 *The effect of friction on free vibration.* (a) *When the frictional forces are predominantly viscous (proportional to velocity), and* (b) *when they are largely independent of velocity.*

Naturally, the more the damping, the faster the vibration decays and one might suspect that damping forces also reduce the natural frequency. This is often but not always true. If the damping forces depend on the *velocities* of the moving parts, they do reduce the frequency somewhat, but fortunately for the theorist the effect is often small in practice. This result is of great value, because it allows accurate estimates of the natural frequency to be made without reference to damping at all. A simple experiment will demonstrate, as illustrated in Fig. 3.7.

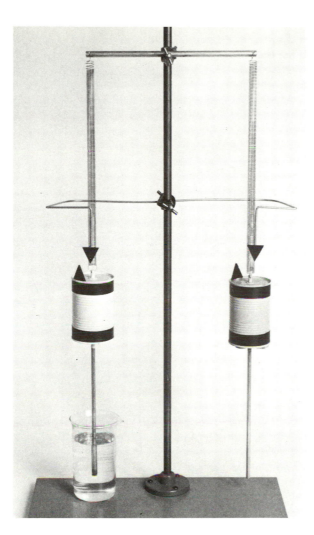

Fig. 3.7 *Free torsional vibration with viscous damping. The two bodies were released simultaneously with equal initial amplitudes. The photograph, taken after several cycles of vibration, shows that, although the body on the left was more heavily damped, it reached the end of the current cycle at almost exactly the same time as the less damped body on the right.*

Two separate and identical spring-mass systems were started simultaneously in free *torsional* vibration. Each body executed a pure angular motion as its spring twisted and untwisted, a vibration similar in kind to an up-and-down vibration but dependent on the stiffness of the spring in torsion and the moment of inertia of the body about the vertical. They kept in step one with the other, confirming that their natural frequencies were equal. A glass beaker containing glycerine was placed below one of the bodies such that the steel rod attached to it was partially immersed in the glycerine. This increased the damping many-fold. When both bodies were once again set into vibration simultaneously, an interesting result was observed. The more heavily damped vibration died down rapidly, but it reached the end of its swings at almost exactly the same instant as the other, right up to the end of the run. This showed that the frequencies of the two vibrations remained very close to each other; the slight drop in the heavily damped one could hardly be observed.

Fig. 3.8 *Natural modes of a two-body system. In the first mode (a), the two bodies vibrate in the same direction at a low frequency; in the second mode (b) they vibrate in opposition at a high frequency.*

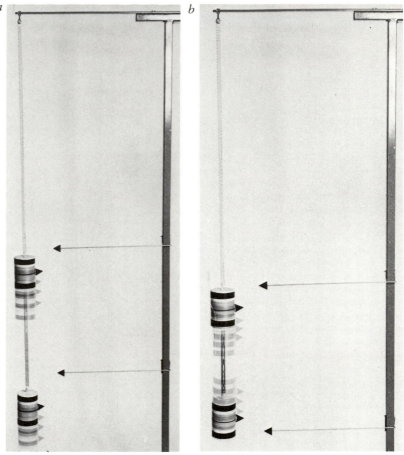

If to the simple spring–mass arrangement of Fig. 3.5, a second spring–mass is added, as in Fig. 3.8, we can discover another important result. Instead of a single natural frequency, *two* natural frequencies will be found to exist, each corresponding to a definite ratio of the two amplitudes. Provided the initial displacements are started off in the correct ratio, the two bodies will continue to vibrate, cycle after cycle, in the same pattern as at the start. Such a pattern is called a natural *mode* of vibration. If started off in some other way, the motion is more complicated, although a careful scrutiny of the result will show that it consists of a combination of the two natural modes.

When the two bodies were displaced by hand by amounts corresponding to the first mode (upper body down 10 cm, lower body down 16 cm*) they vibrated in perfect step with each other at a frequency of 0.62 Hz. To obtain the second mode, the upper body was pulled down 10 cm, and the lower body *up* 6 cm*; once again they carried out a simple vibration, although this time their motions were opposed and the frequency was 1.62 Hz.

When two natural frequencies are close together, peculiar things happen. In Fig. 3.9, the single body suspended on the spring could

* or other displacements in the same ratio.

a
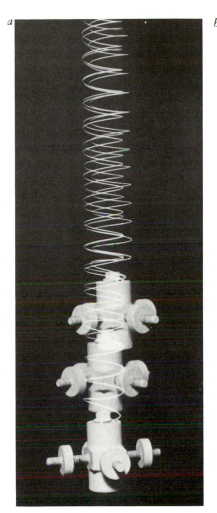

b
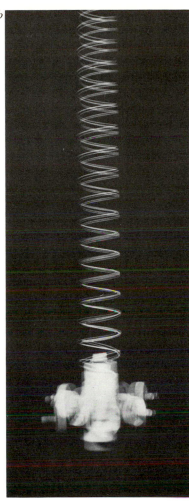

Fig. 3.9 *Vertical and torsional vibrations of the system are coupled through the elastic charcteristics of the spring. If the frequencies of the two types of vibration are nearly equal, a gradual transfer from vertical to torsional vibration takes place: (a) shows the predominantly vertical vibration which gradually transforms into (b), the predominantly torsional vibration.*

vibrate not only up and down but also in angular motion; moreover, the two natural frequencies had been made almost equal. The resulting motion was curious. When the body was pulled down and released, it first vibrated up and down, but soon this motion was accompanied by an increasing angular vibration. In time, the vertical motion virtually disappeared, leaving the angular motion in possession. Then the process gradually reversed until once again the vibration was almost entirely vertical, and the exchange between vertical and angular motion continued until damping stopped the motion altogether. This curious behaviour depends on a little-known property of an open-coiled spring: if it is pulled down by a purely vertical force, it will untwist slightly and, if a pure twisting couple is applied, it will contract slightly. This coupling ensures that in free motion vertical displacements are always accompanied by angular displacements (and conversely). If the frequency of the predominantly vertical mode is nearly equal to that of the predominantly angular mode, each will wax and wane rather like the beats that are heard when two notes close in frequency are played together.

a 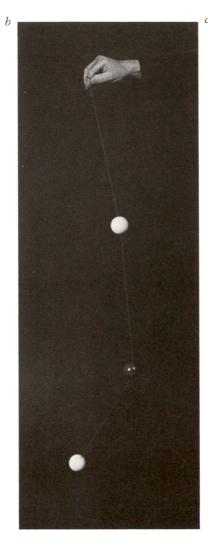 *b*   *c*

Fig. 3.10 *Three modes of conical whirl of a three-body pendulum, similar to the three modes of sideways vibration.*

As might be expected, a system having three spring-supported bodies has three natural frequencies; four bodies, four frequencies and so on. Because it is difficult to start more than two bodies vibrating in a pure natural mode, an alternative but similar arrangement, shown in Fig. 3.10, was adopted. It consisted of three billiard balls suspended at intervals on a length of string and capable of swinging like a pendulum fixed at the top. The spring forces, as in a simple pendulum, were provided by gravity; but, instead of allowing the balls to swing to and fro in a fixed vertical plane, they were set in motion as a *conical* pendulum, each ball following a horizontal circle. To start things going, the top of the string was moved by hand in a small circular motion at a particular frequency, and through a quick process of trial and error a frequency was soon found at which the three balls moved in horizontal circles in interesting and stable patterns. *Three* such frequencies were found, each corresponding to a different pattern or mode of motion. Of course, a conical pendulum differs from a plane or ordinary pendulum, but the mechanics of their

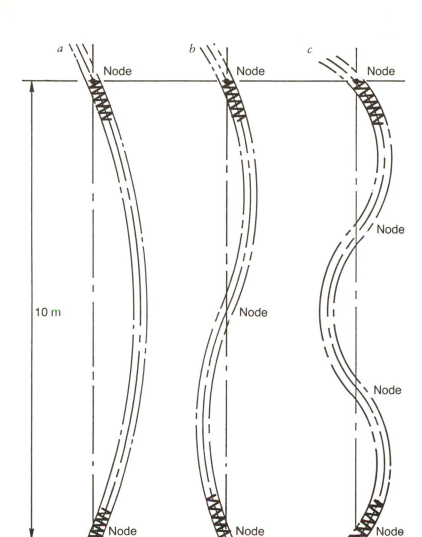

a

Node

10 m

Node

b

Node

Node

Node

c

Node

Node

Node

Node

Node

Node

Fig. 3.11 *Modes of free transverse vibration of a long flexible spring.*

motions are very similar.

Any real machine or structure has both mass and stiffness distributed continuously throughout its parts. In principle, its vibration may be considered as if it had a very large number of discrete small bodies and springs all connected together, sometimes in a complicated way, and with the aid of computer-based methods of analysis this approach has become a practical possibility. Alternatively, if the part under consideration is uniform, such as a shaft of constant diameter, solutions for vibration characteristics can be found by incorporating continuous mass and stiffness in the mathematics at the beginning. In theory, continuous bodies possess an infinite number of natural frequencies and modes, many of which are also important in practice. A good example is a long open-coiled spring, as illustrated in Fig. 3.11.

When such a spring was hung from the ceiling of the lecture theatre it measured about 10 m from top to bottom. It could vibrate freely in all sorts of ways. To begin, it was given a transverse vibration by moving the bottom end slowly from side to side a few times and then holding it still. The spring then vibrated sideways at a low frequency (about 0.4 Hz) in its first mode of transverse vibration – a smooth curve consisting of a single standing wave between top and bottom. When the initial movements had a higher frequency (about 0.8 Hz), the vibration mode was equally stable, but now consisted of *two* standing waves, the upper wave moving to the right as the lower moved to the left. A point near the middle of the spring, called a *node* of the vibration, remained stationary. Several higher frequency modes having increasing numbers of waves and nodes were easily demonstrated.

When, instead of moving the bottom of the spring sideways, the guiding hand moved it up and down a few times to start things going, an entirely different kind of free vibration – longitudinal vibration – was called into play. Like the sideways variety, it also responded stably in a series of compression–extension modes, according to the frequency of the initial input. Another interesting effect was noticed, too. When the bottom of the spring was given just *one* sharp upwards movement and then released, it stayed put for a second or two as a *travelling* wave of compression travelled up the entire length of the spring. When the wave reached the top, it was reflected downwards until it again arrived at the bottom, and only then did the bottom end move.

Now any free motion of an elastic body can be described in terms of its natural modes of vibration, but here we can imagine a different approach based on the idea of a travelling wave. Such a wave is transmitted through the body and may be reflected from a distant boundary if the damping is sufficiently small. Everyone has seen the ripples on the surface of a pond travelling outwards when a stone is dropped into the water; if the edges of the pond are reflective, as with a concrete surround, reflected waves may also be observed. A similar shock wave is transmitted through the ground as a result of an earthquake or an explosion, and we begin to see that we are on the edge of a large and important subject shared by geologists, engineers and others. If we extend the idea to think in terms of a *repeated* succession of inputs rather than just one, and include acoustic and electromagnetic wave propagation as used in submarine detection and radar, the application of wave theory assumes a highly practical significance.

A simple but interesting illustration of a single travelling wave was demonstrated by flicking the bottom of the spring sideways once. A transverse wave travelled rapidly to the top and was then reflected downwards. As it left the top, a second wave was sent up from the bottom. It met the downcomer at about the half-way mark, the two passed through each other with a rapid convulsion and then continued on their separate ways.

Forced vibration

Free vibrations occur when a body supported on a spring is disturbed and then left to itself. The vibration usually dies away because of damping. But many machines and structures have to operate in conditions where a repeated and periodic disturbance is inevitable. A car engine, for example, is driven by a succession of gas forces developed in its cylinders, as illustrated in Fig. 3.4. A rotating machine such as the turbine rotor of Fig. 1.1(c) can never be balanced *exactly*, and the residual unbalance will inevitably exert a rotating force on the machine. It is important to understand the effect of such forces, and much experience has been accumulated as to how to deal with it. This type of motion is known as *forced* vibration.

A familiar example was demonstrated by a volunteer bouncing up and down on a springy wooden plank about 3 m long. At first, a single push downwards showed that the frequency of free vibration of volunteer-plus-plank was about 0.5 Hz. When he raised and lowered both arms continuously at a low frequency of about 0.1 Hz, the plank vibrated in sympathy but with only a small amplitude. A quicker raising and lowering of the arms at 0.2 or 0.3 Hz produced a correspondingly higher frequency vibration, but still with a small amplitude. But when the demonstrator, entering into the spirit of the thing, moved his arms and body at the natural frequency of free vibration – which he quickly sensed by trial and error – the plank responded so exuberantly that he soon bounced off it altogether. Dangerous vibrations of this kind can occur in bridges subjected to continuing periodic forces, arising for example from soldiers marching across in step; bridges have been known to collapse with fatal results and the usual practice is to break step if danger is suspected.

Another familiar but more subtle example of large forced vibration was demonstrated by a swing suspended from the ceiling. The volunteer was given a sharp push to start him going. When he held himself in a fixed position relative to the swing, friction and air resistance gradually decreased his amplitude, but when he leaned backwards on the forward swing and forwards on the backward swing, as everyone learns to do at an early age, he successfully built up the vibration – although, owing to the length of the ropes, he found it hard going. As with many apparently simple motions, the explanation is not so simple. But the chief feature of the leaning action was to provide an offset force by shifting the centre of gravity of the swinger backwards and forwards relative to the seat. This delivered an additional periodic push on the swing at the same frequency as the free vibration, and the vibration built up.

As illustrated by the plank and the swing, the most striking feature of the response in forced vibration occurs when the force causing the motion has the same frequency as the system's natural frequency. This condition is called resonance and it is likely to cause dangerously large vibrations. More generally, the way in which the amplitude of a vibration depends on the frequency of the excitation is an important subject that forms the basis of much vibration testing. If an actuator is attached to a body so as to deliver a periodic force of adjustable frequency at that point, measurements of the resulting amplitudes at other points, taken over a range of input frequencies, can provide much useful information.

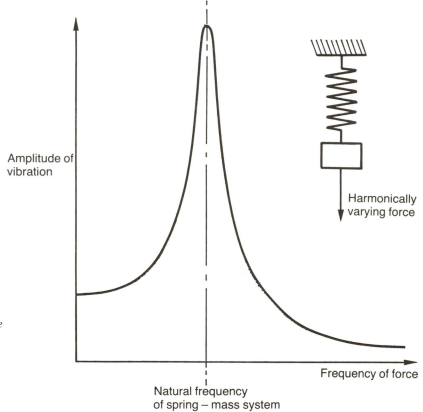

Fig. 3.12 *The response of a simple spring–mass system to excitation by a harmonically varying force. Large amplitudes of vibration occur when the frequency of the force is close to the natural frequency of the system.*

Fig. 3.12 shows how the amplitude of a simple spring–mass system varies with the frequency of an applied force. To interpret the result, imagine a single experiment in which the force varies smoothly with time at a particular frequency such as in the harmonic variations of Fig. 3.3(*b*). If free vibration of the mass is neglected – with damping present, it would soon disappear – the mass will be found to vibrate at the same frequency as the force and at a particular amplitude. In the next experiment, the procedure is repeated; the amplitude of the force is kept the same, but it is given a different frequency and the amplitude of the vibration will be found to be different, too. When, after a succession of similar experiments, the vibration amplitude is plotted against its frequency, the result takes the form of the curve shown in the diagram. The resonance peak occurs near the natural frequency of *free* vibration, when the force varies in step with the system's own vibration characteristic, and the height of the peak reveals the amount of damping present, a high sharp peak

a

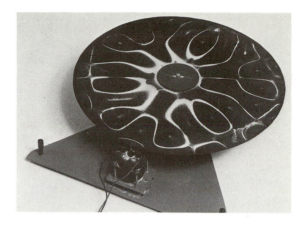

b

corresponding to low damping and a low flat peak to high damping. Practical structures and machines are of course more complicated than this; if, as is usual, a large number of natural frequencies and modes exist, the response curve will contain a number of resonance peaks and the shape of each will depend on the damping in that mode.

Fig. 3.13 shows two interesting types of forced vibration at resonance. The first illustrates bending vibrations of a thin circular steel plate, fixed at its centre and driven by a small electromagnetic vibrator attached near its edge. The vibrator was connected electrically to an oscillator supplying it with a harmonically varying voltage at a frequency controlled by the operator. When the input frequency was made equal to one of the many natural frequencies of the plate, a large and noisy vibration ensued, which was made visible by sprinkling a liberal dose of dry salt on the surface of the plate. The results for different modes are shown. As the plate vibrated, it bounced the salt off the moving parts to the areas where little or no motion occurred, leaving a distinct pattern of nodal (or stationary) lines clearly displayed. Some, like the butterfly mode, looked simple, and others complicated.

In the second experiment, a long closed metal tube was supplied with inflammable gas which could escape upwards through a number of small holes drilled along the length of the tube. As the gas escaped, it was ignited, and a line of small flames, roughly equal in height, was

Fig. 3.13 *Forced vibrations at resonance. (a) Bending vibration of a thin steel plate fixed at its centre: various patterns of vibration corresponding to the natural modes of free vibration are made visible by loose particles of salt settling on the nodal lines. (b) The height of the flames reveals the pattern of longitudinal vibration of the inflammable gas in the tube.*

seen to be issuing forth. A microphone, fitted to one end of the tube and connected to a variable frequency oscillator, put a vibration into the enclosed column of gas, which thereupon vibrated longitudinally rather like a long spring. In turn, the motion of the gas column affected the pressure of the gas escaping from the holes, and the height of a flame – which depended on the average pressure at that point – varied along the length of the tube. When, by adjustment of the microphone's frequency, a resonance was reached, a pronounced pattern of vibration could readily be seen – a single standing wave at the lowest frequency and progressively larger numbers at higher frequencies.

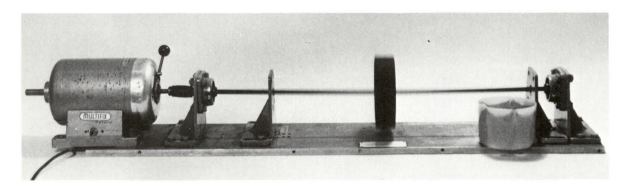

Fig. 3.14 *Whirling vibration of a flexible steel shaft when running at a critical speed.*

Forced vibration often occurs in rotating machines as a result of unbalance. In Fig. 3.14 a long steel shaft is shown supporting a single rotor at its centre. When the shaft was still stationary, the rotor was given a sharp sideways tap, at which the shaft performed a free bending vibration at the natural frequency of just over 11 Hz. As the assembly was slowly driven up to speed, it responded to the excitation of the inevitable rotating unbalance by executing a forced whirling vibration at the same frequency as the shaft speed. At first, the amplitudes were small, but as the speed approached 700 rev/min (11.7 Hz) the motion became violent. Some water in a small beaker standing on the base of the apparatus joined in the affair with much agitation until, as the shaft continued to accelerate, the vibration ceased and peace was restored. At very high speeds everything ran smoothly again. This kind of resonance is aptly known as a critical speed, and one of the most important requirements in the design of rotating machines is to ensure that they do not run continuously at a critical speed. When passage through a critical speed is inevitable, as in large turbine generators, the passage must be quick enough to prevent vibration growing to serious levels. Similar problems occur in car engines, where the crank-shaft and transmission shafts are susceptible to *torsional* vibration, and, because car engines are required to operate over a wide range of speeds, the suppression of resonance is an important aspect of their design.

In theory, the simplest method of eliminating a forced vibration is to eliminate the forces which cause it, but unfortunately this approach rarely offers a practical solution. Disturbances should obviously be reduced as much as possible, but that may not be enough. The design of the machine must ensure that all significant natural frequencies of free vibration are well removed from the frequencies of any disturbances; such design calculations often require extensive computer-aided procedures to explore the effects of changing masses and stiffnesses.

Other methods which reduce forced vibration include the addition of damping. But because damping may also reduce the performance of the machine to unacceptable levels, it cannot be invoked indiscriminately. Vibration absorbers, consisting essentially of a mass judiciously attached to the machine by means of a spring, offer useful possibilities for certain applications; if the natural frequency of the absorber is made the same as the frequency of the disturbance, the absorber's own vibration will help to suppress the vibration of the parent machine.

Neither free nor forced

Not all vibrations fall conveniently into free or forced categories. From the point of view of the vibration specialist, many of the more interesting varieties are of quite a different kind and, as we shall see, they also have important practical implications.

To illustrate the matter, the model shown in Fig. 3.15 was put through its paces. It consisted of a V-section channel of aluminium, balanced on a knife-edge at its centre, and held in a horizontal position by a small weight acting as a pendulum. A thin dividing plate separated the left-hand and right-hand lengths of the channel. When water was released from a vessel mounted above the divider, about half of it went into the left-hand side to pour off the left-hand end, and the rest went to the right. Inevitably, the flow was not divided exactly in half and slightly more water went to one side than the other. This caused the heavy side to move downwards. As it did so, the divider deflected more water to the other side and soon the position was reversed; the up-side overbalanced the down-side and the channel swung the other way. Once again, more water was diverted the other way, the cycle of events was repeated and soon a steady vibration was in full swing.

This simple example shows how a vibration can develop without any external periodic force whatever. The vibration itself generates the fluctuation in the force, which in turn increases the vibration, which further increases the force, and so on until something happens to limit the motion – damping or perhaps destruction. It is known as a self-induced vibration, and examples can be found in many walks of life.

Probably the humblest is the rubbing of a forefinger along a fairly smooth table top. For the same downwards pressure, the frictional force bending the finger goes *up* slightly as the forward speed of the fingertip goes *down* and if the finger deflects backwards by a small amount, it receives an extra push forwards which causes it to

Fig. 3.15 *A self-induced vibration.*

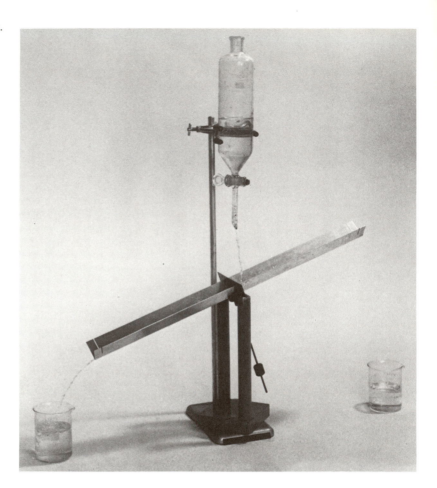

overshoot its mid-position relative to the hand. As it catches up and overtakes that position, its speed is slightly higher than average, which in turn reduces the frictional force and once again it lags behind. All this happens very quickly, but it can produce a satisfying buzzing noise, as many young experimentalists have discovered for themselves.

Something similar happens when a violin is played, Fig. 3.16. The initial movement of the bow deflects the string slightly and because the bow at first moves in the same direction as the string, the frictional force between them is slightly more than its average value. At a certain deflection, the tension in the string overcomes the friction of the bow, the string moves back, the friction falls slightly below its average value and the conditions are ideal for a sustained vibration. The string vibrates at its natural frequency, which depends on its tension, its mass and its length, and to get the right note the violinist must ensure that the tension is correct before he starts to play.

Self-induced vibration can also make an unwelcome appearance in rotating machinery. Sometimes the road wheels of cars develop an annoying vibration which plays back into the steering wheel; although many such troubles are probably due to a forced vibration caused by unbalance in the wheels, genuine self-induced vibrations

Fig. 3.16 *Another self-induced vibration.*

a

b

Fig. 3.17 *A model demonstrating self-induced vibration of railway wheels. Courtesy of British Rail.*

have occurred in road vehicles, aircraft nose-gear and railway wheels. Fig. 3.17 shows a model of a pair of railway wheels supported on a pair of discs representing the rails. To simulate the action the discs are driven by a variable-speed electric motor and, in turn, drive the wheels by friction. The axle to which the wheels are attached is held in a central position by a springy steel rod representing the suspension of an actual wheel-set. In normal running, the spring helps to keep the wheels central; but, if for some reason they should move off centre, a reliable steering control is provided automatically by their geometry. Unlike a simple cylinder, the edges of the wheels have a slight taper – a larger radius on the inside and a smaller on the outside. If they move off-centre to the left as viewed in (*a*), the left-hand wheel will run on a larger radius than its companion on the right. Because they both revolve at the same speed, the left-hand end of the axle moves forward

faster and so turns the combination back to centre. So far, so good –
but unfortunately that is not the end of the story. Under certain
conditions, complicated dynamical effects associated with the mass of
the wheels, friction and the spring suspension can carry the wheels
past the central position marginally further than the original
departure, and the cycle of events is repeated on the other side. A self-
induced vibration then develops which may cause clashing of the
wheel's flanges against the inside of the rails. This phenomenon has
been investigated extensively by railway engineers, who have worked
out the details of the mechanics and established the correct rules for
design.

The model was first run at a low speed and the wheels given a sharp
blow sideways. They moved to and fro a few times in a free vibration
then settled down to smooth central running. As the speed of the drive
was gradually increased, the sideways motion became more lively
until at yet higher speeds, a full-blown vibration developed without
any external encouragement whatever. It grew progressively larger
until the wheel-pair hit the stops on either side. When the speed was
reduced, central running was restored.

Many self-induced vibrations have been caused by fluid or gas
flows. In the 1930s, as the speeds of aircraft increased and their
structures became more efficient, a new phenomenon now known as
classical flutter began to plague aeronautical engineers. It consisted of
a combined bending and twisting of the wings and depended on the
way in which the aerodynamic lift force varied with these deflections.
Now that wings are often more like plates than beams, the
calculations have become more difficult, but it is still essential at the
design stage to ensure that the airspeeds at which flutter will occur are
well above the operational limits of the aircraft.

Other self-induced vibrations have been experienced in a wide
variety of structures – metal chimney stacks, submarine periscopes,
boiler-tubes and bridges. The precise mechanics in these cases can be
difficult to establish, especially when both resonance and self-induced
vibration are involved. If a solid body is placed in an airstream, the
flow of air is split into two streams which recombine in the wake. In
the region of recombination, the two streams have to adjust
themselves to their different velocities, and they can do so by forming
vortices in a regular pattern, first on one side of the mean line of flow
and then on the other, as shown in Fig. 3.18. These vortices are
accompanied by regular pressure variations on the surface of the body
at right angles to the direction of flow, thereby generating a
fluctuating force at a particular frequency. If the structure is capable
of vibrating in the direction of the force, and if the frequency of the
force is close to a natural frequency of the structure, the stage is set for
large and possibly destructive vibrations to appear. Moreover, the
frequency at which the vortices are shed is influenced by the vibration
itself; if the result is coincidence, the structure is in serious trouble.

Another kind of self-induced vibration was illustrated by the light
balsa-wood beam of semi-circular cross-section shown in Fig. 3.19. It
could bounce up and down on the springs connecting it to the frame,

Fig. 3.18 *Vortices shed from a blunt body drawn through still air. From Hertel, 1966. Courtesy of Reinhold.*

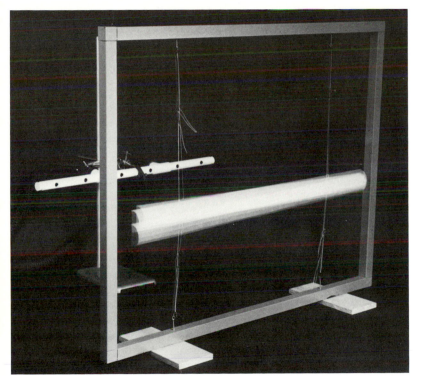

Fig. 3.19 *A modest flow of air supplied through the holes in the fixed tube caused a destructive vertical vibration of the spring-mounted beam.*

and because it was so light the forces exerted upon it by the air through which it moved played a significant part in the proceedings; as the beam moved upwards, it experienced a downwards force from the air and vice versa. When a modest stream of air was blown over the beam, the beam first moved slightly in the direction of the airflow; but, as the airspeed was increased, a remarkable *vertical* vibration began – at right angles to the airflow – and grew to such an extent that one of the supporting springs snapped loose. Clearly, despite the steadiness of the wind from the blower, a pulsating force was at work that was created by the vibration itself. Any slight movement of the beam *upwards* from rest had the effect of introducing a *downwards* component of the air flow relative to the beam, with the result that as the airstream divided over the beam the flow travelled slightly faster over the top edge than over the bottom. According to the principles of fluid flows, this *reduced* the pressure at the top compared with that at the bottom, and a net upwards force was generated, pushing the beam further up. However, the spring forces eventually brought that movement to a halt and the beam began its descent, with similar results except that now the net pressure force was downwards. The fluctuations in this vertical force were enough to build up the vibration to destruction.

Fig. 3.20 *A model bridge-deck supported on springs developed a large vertical vibration as air was blown over it.*

In the second experiment, the section of the beam was more like a bridge structure, with side-faces and a bridge-deck (Fig. 3.20). Once again, as air was blown over the bridge, a convincing vibration developed – although this time it was stopped before everything fell to pieces. As well as the aerodynamic self-excitation of the previous example, the detailed mechanics now also included vortices shed alternately from the sharp top and bottom edges of the leading side-face. Together they caused vertical pulsations in the air pressure on the beam, and at a certain wind speed the frequency of the pulsations approached one of the natural frequencies of the beam on its spring.

Such a vibration led to the destruction of a large suspension bridge crossing the Tacoma Narrows in the State of Washington (Fig. 3.21). The bridge was much more complicated than our simple model. It

Fig. 3.21 *The last moments of the Tacoma Narrows bridge, 7th November, 1940.*

could vibrate in several bending and torsional modes, all at different frequencies, and it began to exhibit unusual and persistent vibrations soon after it was opened on 1st July, 1940. Nevertheless, traffic was allowed to use it for several months until by November the vibrations had become so severe that it was closed. By that time it had attracted widespread attention. Measurements were taken, records kept and, finally, a dramatic film was made of its final collapse on 7th November, 1940. Intensive investigations followed which threw much light on the aerodynamics of bridges and its interplay with structural vibration.

Another kind of vibration can arise in machines when an internal feature of the parts, such as a spring stiffness, varies periodically with time in a pre-ordained way. Such arrangements are known as *time-dependent* systems, not because their motions in vibration depend on time (which is of course true for any vibrating part) but because their own inherent characteristics vary with time as well. The combination leads to entirely new phenomena, of which the most striking is an instability of a spectacular kind.

A simple model showing this kind of instability consisted of a pointer that could swing freely as a pendulum from a support near its upper end, Fig. 3.22. Ordinarily, of course, it behaved in a perfectly credible way; when moved to one side, it oscillated about its position of equilibrium. But there was more to the model than met the eye. The point of support could be vibrated vertically by means of a linkage connected to a variable-speed electric motor. This had the effect of varying the effective gravitational field controlling the swing of the pendulum – in other words, the spring force was made to vary at the same frequency as the vertical vibration of the support. When that frequency was low, the pointer behaved much as an ordinary pendulum. However, as the frequency was gradually increased, the excursions from a downward position became increasingly lively, and a frequency was soon reached at which the pointer became distinctly unstable. At yet higher frequencies it left the downwards position altogether and proceeded to oscillate about an upwards position, even when pushed firmly to one side.

Fig. 3.22 *An unusual pendulum.
When the pivot is stationary as in
(a), the pendulum oscillates in the
ordinary way about the
downward position; but when the
pivot is given a vertical oscillation
(by means of a motor behind the
board) the downward position
becomes unstable under certain
conditions, and instead the
pendulum oscillates about an*
upward *position, as in (b). The
conditions governing downward
instability are shown in the
stability chart of (c).*

a 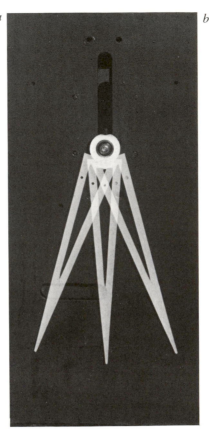 *b*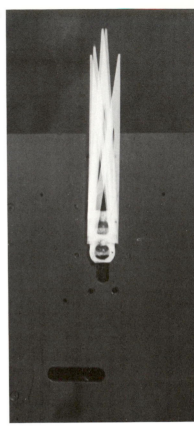

An explanation of such eccentric behaviour takes us into the further
reaches of applied mathematics, where we find that stability depends
on the variation of the effective spring stiffness. Not all arrangements
can offer a *second* position for stable oscillations, and usually the
most important question concerns the stability of the first or normal
kind of oscillation. The stability chart shown in Fig. 3.22(*c*) deals with
this question. It shows that certain combinations of pivot frequency
(in relation to the pendulum's natural frequency) and amplitude of the
pivot's acceleration (in relation to **g**) give rise to instability, as
indicated by the shaded areas, whereas outside these areas the
downward position remains stable. The boundaries, although nearly
straight, become decidedly more complicated outside the small corner
of the chart illustrated. Outside such ranges, the downward position is
stable. This is strange but important behaviour not only for the
pendulum but also for machines having similar time-dependent
characteristics, such as a shaft having a section like that of Fig. 3.23.

To simplify matters let us suppose that the shaft is supported in
fixed bearings at its ends and that it is constrained by guides to deflect
in the fixed direction *OX* only. It is easy to see that if it were free to
bend in any direction the shaft would be more difficult to bend in
direction *Oy* than in direction *Ox*. It follows that as the shaft rotates,
it will offer a variable bending stiffness to deflections in direction *OX*:

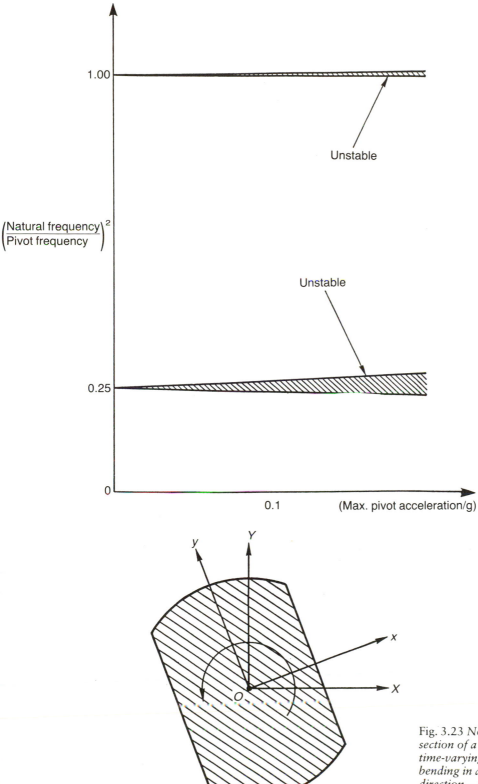

Fig. 3.23 *Non-circular cross-section of a rotating shaft offering time-varying stiffnesses to bending in a fixed transverse direction – and a potential for unstable vibration.*

in other words, it constitutes a body with a time-dependent stiffness. Many shafts used in machines, such as those in which longitudinal slots have been cut for various purposes, possess this characteristic to a greater or lesser degree. Their bending stiffnesses vary twice per revolution, and they can display the same kind of instability as the pendulum.

Fig. 3.24 *Torsional vibration of a familiar time-dependent system: well-timed pulls on the string produce a satisfying, reversible and fast spin.*

An easily constructed model, once a popular toy, consists of a thin cardboard disc (Fig. 3.24) through holes in which a loop of string has been threaded. If the string is held in a more or less horizontal position and given a slight twist to begin the motion, the disc will respond to a pull on the loop by spinning in a direction that untwists the string. By judicious timing of a succession of pulls, a convincing oscillation can be generated in which the string first unwinds, then winds up the other way and so on, giving the disc a fast oscillatory rotation while emitting an interesting noise. The time-dependent feature is the tension in the string, which controls the twisting or untwisting torque on the disc, and the limit of the motion is determined by the ability of the operator to withstand the nip the winding string exerts on the fingers. More thoughtful designs, however, provide small metal rings at the ends of the loop to eliminate this hazard.

A related but different kind of vibration forms the basis of mechanical clocks and watches. The history of these machines from ancient times not only records marvels of human ingenuity but also shows how the craftsmen involved made major contributions to the art of fine mechanical construction. The basic requirements can be stated simply enough – given a driving force provided by a falling weight (or an unwinding spring) how should the motion be controlled such that for

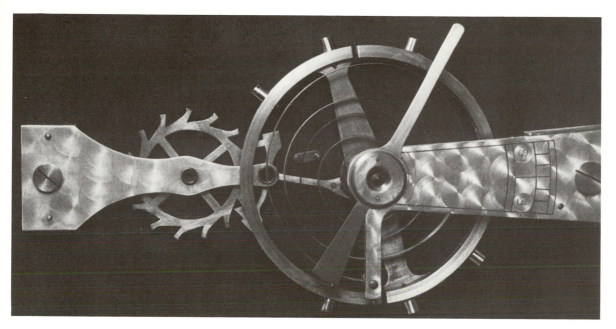

Fig. 3.25 *An enlarged model of a watch mechanism. The small driving wheel on the left is connected to the hands and is driven by the mainspring through gears (not shown). Its motion is interrupted regularly by the fork-shaped lever, which is made to oscillate in step with the oscillations of the balance wheel on its spring on the right.*

a short drop (or unwinding) a hand of the clock will keep turning regularly and accurately for a period of days? Clock makers found many clever solutions to this problem, one of which is shown in the watch mechanism of Fig. 3.25. As the mainspring unwinds, it drives through gears (not shown) a wheel, the motion of which is interrupted regularly by a vital part of the whole assembly, the escapement mechanism. This consists of an oscillating or balance wheel, an attached balance-spring and a lever shaped like a fork with two widely spaced prongs. The frequency of the balance-wheel is crucial to success. As it rocks through a small amplitude it oscillates the lever, and the pallets or prongs at the ends of the lever periodically engage with the driving wheel, first on one side and then on the other, thereby allowing the wheel to move forwards one tooth for each swing of the lever.

The details of the motion are far from simple, but certain observations can readily be made. First, the amplitude of the balance-wheel's oscillation is determined by the geometry of its engagement and disengagement with the driving wheel. Secondly, the driving wheel – and the hand to which it is attached – will both advance in a regular way as required for time-keeping. Thirdly, the force or torque of the mainspring driving the whole mechanism is controlled by the regular action of engagement and disengagement – it cannot simply cause the hand to run away. And finally (this is where the details are crucial), several quite different phases of motion occur in each and every cycle of engagement and disengagement, according to whether the principal parts are in contact or not. In other words, the mechanism as a whole changes its internal characteristics as each cycle proceeds. Unlike a time-dependent system, however, the changes depend on the rearrangements of the parts caused by the vibration itself, and we find ourselves in the territory of self-induced vibrations of a special kind.

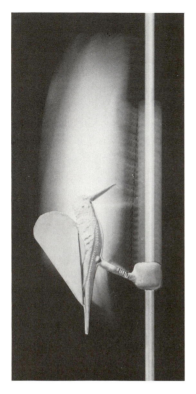

Fig. 3.26 *A popular toy of days gone by. Once started, the vibration of the bird on the spring connecting it to the collar releases the friction holding the collar on the rod, and a jerky descent ensues.*

Two amusing toys also depend on the way in which the moving body rearranges itself so as to maintain a repetitive motion. Fig. 3.26(*a*) shows a wooden model of a bird that can slide down a fixed metal rod. It is connected to the rod via a small spring, the other end of which is attached to a sleeve fitted loosely around the rod. If left to itself, the weight of the bird will bend the spring downwards and the sleeve will be locked by friction in a slightly tilted position. When the bird is given an upwards tap, it begins to oscillate on the spring, which in turn releases the friction lock during the upwards swing. The sleeve obligingly slides down the rod for a short distance until the next downwards swing of the bird once again locks the sleeve on the rod. By this means the bird gradually descends to the bottom in a series of oscillatory jerks, and if its beak is made long enough to touch the rod at the end of each upwards swing the descent will be accompanied by a succession of staccato taps.

The other model, Slinky the spring, is best known for an ability to travel downstairs in a series of hops, as shown in Fig. 3.27. Once again, the motion is driven by gravity, and if Slinky is released from a thoroughly bent position, with the top coil held on one stair-tread and the bottom coil on the next, the spring will first straighten out, overshoot the mark, topple over until the top coil hits the next tread and, provided the spring is well-matched to the stairs, continue regularly to the bottom. Both Slinky and the bird succeed in rearranging themselves in the course of a single cycle to take advantage of the steady gravitational force.

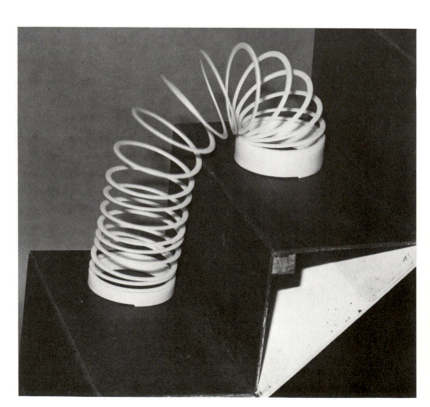

Fig. 3.27 *Slinky going downstairs.*

4

UNDER CONTROL

Controls for everything

The safe and accurate control of machines constantly growing in speed and power is obviously a matter of some importance; a machine that regularly runs amok is of very little use to anyone. And, although control theories were first developed for machines, their central ideas find application in many other walks of life as well. We have only to think of the control of processes (one of the earliest automatic controllers, a weight resting on a hole, was used to regulate the pressure in a primitive pressure-cooker), traffic control, the control of our own movements, and even attempts to control economic cycles to realize that we are dealing with a subject of practically universal application.

Let us consider something controlled by a fixed set of instructions. Traffic lights follow a sequence of green, amber, red, red and amber, green . . ., and thereby control the flow of traffic at an intersection. Except for those cities where computer-based methods are available, the interval for each colour is pre-set and independent of the volume of traffic. Many simple machine controls operate in a similar way; most washing-machines, for example, run through a fixed cycle once the controls are set by the operator. Fixed programmes of this kind have a long and interesting history, including the automata that were all the rage in Europe about 200 years ago. The still popular music box recalls the melodies of an earlier age, and in Fig. 4.1 the famous silver swan of Bowes Museum illustrates one of the most elaborate artefacts of that era. Such devices were the forerunners of numerical

Fig. 4.1 *The Bowes silver swan: this beautiful example of a mechanical automaton, driven by clockwork, periodically immerses its beak in a pool of water and appears to swallow (artificial) goldfish. Courtesy of Bowes Museum.*

control in industrial processes. Now, almost two centuries later, the same principles are being applied via computers to lead the revolution in manufacturing that is making such dramatic progress.

Extensive theories have been developed to account for the behaviour of machines operating under one kind of control or another, and the results have led to spectacular gains in design and performance. Such analyses have to accommodate types of control that are as varied as the systems upon which they operate. Some, like traffic lights, work on a fixed programme; others, known as regulators, are intended to keep some desired quantity constant, such as the speed of a rotating machine or the flight-path of an aircraft. Many control systems operate continuously, as in a servomechanism controlling the angular position of a remote shaft, while others, often used in chemical processes, employ a controller which samples the results at intervals. Some controllers even adapt themselves internally so as to improve their own performance.

Inevitably, the range of equipment involved is very wide. It includes instruments for measurement, actuators for generating forces and computers for processing information. Fortunately, however, when all are assembled, the resulting systems share certain common principles.

Closing the loop

The idea of a closed-loop is so important in control systems that it is useful to consider first a different and simpler mode of operation. As we shall see, this alternative, known as open-loop, has severe limitations.

Suppose we wish to run a d.c. electric motor driving a disc at a certain speed. We know that with a particular electrical construction

Fig. 4.2 Open-loop speed control: (a) the setting of the potentiometer controls the speed of the disc, but any change in the load torque affects the result; (b) the corresponding block diagram.

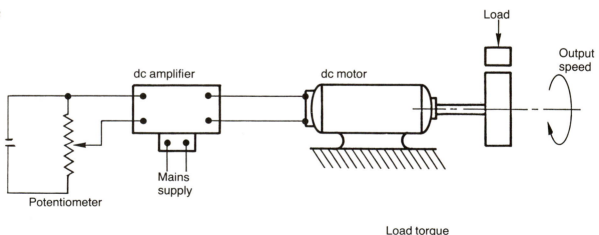

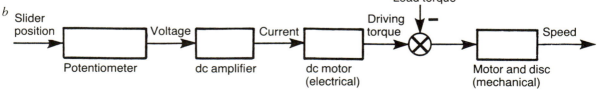

the current flowing through the field windings affects the speed of the machine. Fig. 4.2 shows how the current might be controlled; a potentiometer or sliding resistance supplies a controllable voltage to an amplifier which in turn delivers a proportional curent to the motor. Suppose we find by trial and error the position of the slider that delivers the correct current for the required speed. All would then be in order. But if the load torque on the disc were to change, we should find that the speed would change too; more load, less speed, and conversely. Our initial setting would no longer be effective. Matters would be worse if the motor were required to run at *different* speeds at different times, for, if the positions of the slider had been carefully marked to yield these speeds, all would fail when the load changed.

The picture we naturally form of an open-loop control is illustrated in Fig. 4.2(*b*). The blocks represent the physical components of the system, and the connecting arrows represent the quantities transmitted from one to another. Such a block diagram enables the control engineer to assemble a great deal of information in a very effective way. The electrical characteristics of the motor, for example, can be represented by a mathematical expression written inside the block such that the quantity coming *out* is equal to the quantity going *in* multiplied by the expression (called the transfer function) inside the block. In this way complicated control systems can be broken down into their constituent parts, each of which can be assigned to a suitable transfer function.

A simple example of transfer functions can be illustrated with two steel levers of equal lengths, one stiff, the other flexible, and both operating on the same pivot P as shown in Fig. 4.3. If we ignore the

Fig. 4.3 *Two levers: the output displacement of the stiff lever is proportional to its input, but the relationship is not so simple if the lever is flexible.*

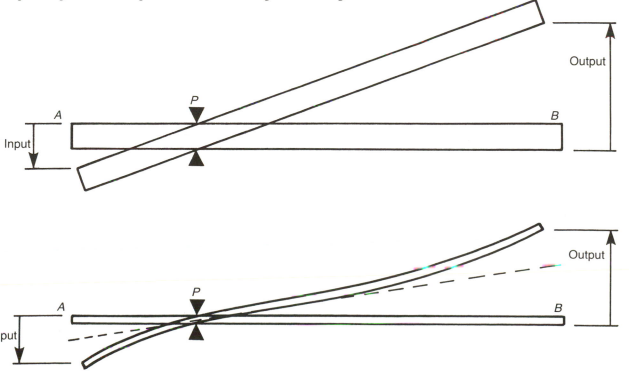

slight bending of the stiff beam as end *A* is moved sharply downwards through a given displacement (the input), the upward displacement of end *B* (the output) will be directly proportional to the input. The transfer function relating output to input is simply the fixed ratio *BP/AP*. When the experiment is repeated with the flexible beam, the upward displacement of *B* is not so straightforward; because of the flexibility, *B* will move past its final position and oscillate before coming to rest. The transfer function relating output to input now has to take the dynamics of the lever into account. Similar dynamical effects occur in almost all control systems, but fortunately transfer function analysis is equal to the task and allows vital dynamical characteristics to be incorporated in the proceedings.

Let us return to the idea of loops. In view of the disadvantages of open-loop control, matters would be much improved if the controller could take account of variations in the load – or for that matter of any inadvertent disturbances at the output. Fortunately, the means are readily to hand – we simply close the loop. In so doing, we change the whole performance of the control system for the better, including its dynamical as well as its steady-state characteristics.

The physical additions required for the motor speed control can be summarized simply. First, we attach a device to the motor-shaft that will measure the actual speed and generate a proportional electrical voltage. This voltage is then compared with the voltage supplied from the input potentiometer, and the *difference* between the two voltages is applied to the amplifier. If the load should now increase, causing a loss in speed, the amplifier current automatically increases to compensate for the loss. Naturally, the position of the input potentiometer must first be calibrated in terms of desired speed, and, as any subsequent variations in load affect this calibration, this simple arrangement will fail to provide perfect compensation. Nevertheless, the residual errors will be greatly reduced, and for good measure the dynamical performance can also be much improved.

The corresponding change in the block diagram is shown in Fig. 4.4, where the loop has now been closed. Because the quantity fed back from the output is *subtracted* from the reference, it is described as *negative* feedback, and as we shall see negative feedback plays a powerful rôle in many of nature's controls as well as in machines.

Fig. 4.4 *Closed-loop block diagram for the speed-control system of Fig. 4.2 (a).*

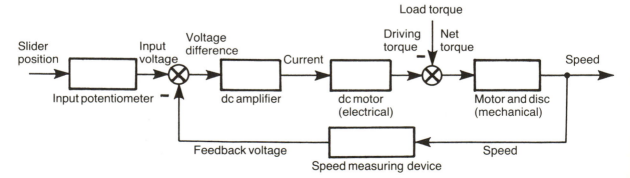

One of the earliest speed controllers for rotating machinery was used by James Watt in the famous steam engines that dominated the production of mechanical power towards the end of the 18th century. The device, the centrifugal speed governor, was purely mechanical and is illustrated in the model of Fig. 2.5(*b*). Its drive-shaft was connected directly to the engine-shaft. If the engine speed increased, the fly-weights and the hinged links would move outwards; the lower links then moved a sleeve up the drive-shaft, which in turn closed the steam supply valve through a connecting-rod so as to reduce the engine speed.

It was soon discovered that under unfavourable circumstances the momentum of the various parts would carry the assembly beyond the desired position. More steam than necessary would be admitted, the engine speed would rise excessively and a cycle of instability, known as hunting, would develop. Mathematically, hunting resembles the self-induced vibrations described in the preceding chapter; it is a condition to which many other control systems endowed with a plentiful supply of energy are also liable. The first comprehensive analysis of this troublesome effect was published by the eminent mathematician James Clerk Maxwell in 1868.

The general features of open-loop and closed-loop control were demonstrated by means of the arrangement shown in Fig. 4.5. It consisted of an electric motor, connected to an amplifier fed by a potentiometer, and driving a disc against friction created by pressing a pad against the edge of the disc. The pad was mounted on the end of a lever such that the friction load could be changed by moving a weight along the lever. For open-loop control, a volunteer was invited to find three positions of the potentiometer slider corresponding to speeds of 50, 100 and 150 rev/min. He did so very expeditiously; but, when the friction load was increased at each of his marked positions, the speed dropped substantially. A second volunteer was then invited to take

Fig. 4.5 Speed control. The position of the small weight on the right determined the frictional load on the disc. The driving motor behind the disc was controlled by the sliding potentiometer via an amplifier in the box. It was the task of the hand on the left to keep the speed as constant as possible despite the unpredictable actions of the hand on the right: although the left hand did not know what the right hand was doing, visual feedback supplied by the speedometer successfully kept excursions within limits.

charge of the load, and to change it as rapidly and unpredictably as he could. By continuously observing the changes of speed displayed on the dial, the first volunteer – the controller – was able to close the loop, the feedback taking the form of his reading of the speed. Whenever he saw the speed differing from the desired value, he adjusted the potentiometer so as to return the speed to its correct value – a human link in the closed-loop. After a few trials, he found he could successfully foil the attempts of the loader to change the speed significantly; but if, as James Watt had demonstrated with his speed governor, the human link had been replaced by an automatic device, the machine could have been left to look after itself.

Where human movements are involved, the ways in which feedback operates can be surprisingly subtle. When we stand upright on two legs, for example, we can never remain perfectly still. Instead, using complicated feedback signals, our muscular controls are continuously called into action to limit the inevitable swaying of our bodies. A simple experiment along these lines is to stand on *one* leg, first with eyes open then with eyes closed. You will find it is not at all easy to stay upright for long without the help of feedback provided by vision.

Another simple experiment consists of supporting a wooden rod, say 1 m long, in an upright position on an outstretched finger. With practice, the rod can be kept upright for a short time provided the owner of the finger moves it skilfully – up and down, front and back or sideways – using the feedback of his or her eyes and the pressure on the finger to determine what should be done next. Surprisingly, the task can be made easier by adding a weight at the top of the rod, when the increase in the moment of inertia of the assembly about the finger more than compensates for the additional toppling moment of the weight.

The long arm of control

A more elaborate experiment was run with a remotely controlled model jeep known as Wild Willie, Fig. 4.6. This small machine was capable of responding to radio signals emitted from a hand-held transmitter, one switch controlling the front-wheel steering and another the forward-and-reverse drive. First, a volunteer, Timothy, was invited to drive the model through a narrow gap between two obstacles, which he skilfully accomplished not only forwards but also backwards. The second and more difficult task involved simulating a delay of just over one second in transmitting a picture from a television camera viewing a real jeep on the surface of the Moon to an observer on Earth. In reality, the Moon-vehicle would also experience a delay in receiving the control signal *from* the Earth, but in the lecture theatre only the one-way delay from Moon to Earth was taken into account.

With much ingenuity, the BBC engineers had arranged for the volunteer to watch a TV screen on which still pictures were displayed showing a succession of positions of the model jeep, all taken a second or so previously; the audience could see *both* the current and the

a *b*

earlier positions, side by side on a screen, but Timothy was denied that advantage. With this in hand, the experiment began. Timothy performed admirably. Carefully watching the out-of-date picture and shrewdly estimating what was happening in reality as he operated the switches, he triumphantly drove the jeep through the gap, to general applause.

During the Second World War, control engineering made rapid advances as necessity gave birth to invention. Applications of the greatest importance required that the angular position of a large object such as a radar antenna or a heavy gun-barrel should be controlled accurately and quickly by means of a small control-wheel in a remote place. Such devices for the control of position became known as servomechanisms, or slave mechanisms. Fig. 4.7 shows a laboratory model (a modified version of the motor speed control system) which worked in the following way.

When the operator turned the input wheel to a desired position, the output shaft followed. The difference between their angular positions was registered as an electrical voltage by means of identical potentiometers attached to the two shafts. This comparison constituted the feedback, as depicted in Fig. 4.7(*b*). The difference or error signal was transmitted to an amplifier, which multiplied the signal before feeding it to the motor terminals. As long as an error persisted, the output shaft was driven to the required position. Such an arrangement provides a rapid and accurate control for large objects, the power being provided by the electric motor. Moreover, the performance of the system can be varied easily by adjusting various settings to meet any particular requirements of speed, power or damping.

Various characteristics were demonstrated by such adjustments. A volunteer first turned the input wheel slowly. The output shaft

Fig. 4.6 *The time-delay experiment. When the actual position of the model jeep was as shown in* (a), *the volunteer controller could see only a television picture of an earlier position shown in* (b). *Despite this disadvantage, he successfully steered it between the obstacles, using a move-and-wait strategy.*

a

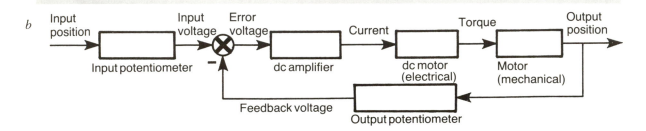

b

Fig. 4.7 *A servo-mechanism.*
(a) *When the hand moved the input lever on the left, the motion was faithfully reproduced by the disc on the right, as explained in the text; with longer wires, the disc would continue to follow instructions even if placed in a remote position. (b) The corresponding block diagram.*

followed closely as shown in curves (i) of Fig. 4.8. Next, he gave the input sheel a sudden turn. This caused the output shaft to overshoot the mark and to perform a few oscillations before settling down as shown in curves (ii) – behaviour that might be very undesirable in practice. The next run showed how the overshoot and oscillation could be reduced. By means of a simple adjustment of the electrical settings, an additional signal was introduced in the feedback, proportional to the *angular velocity* of the output shaft; this effectively introduced damping, and both overshoot and oscillation were abated.

Finally, the volunteer was replaced by a device that oscillated the input in a pre-determined way. At low input frequencies, the output shaft oscillated through a small amplitude, but as the frequency increased the oscillations at output continued to grow until, at still higher frequencies, they once again decreased. This pattern of behaviour is illustrated in Fig. 4.8(*b*), where the amplitude of the output is plotted as a function of frequency. The curve resembles the forced vibration response of a mechanical system, as discussed in the previous chapter, and the mathematics of the two is very similar. For practical purposes, such tests can provide the control engineer with valuable information about the characteristics of complex systems.

Cutting the cost

To anyone faced regularly by heating bills, the prospect of better controls has an instant appeal. If the supply of heat corresponded more closely to demand, resources would be better employed and the cost of keeping warm would be much reduced. In another familiar field, motorists would undoubtedly take a favourable view if the way

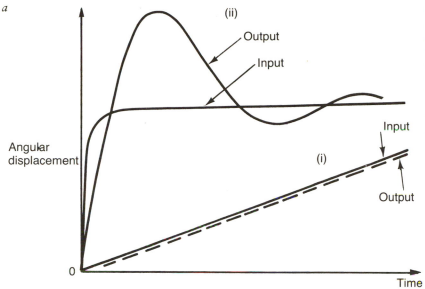

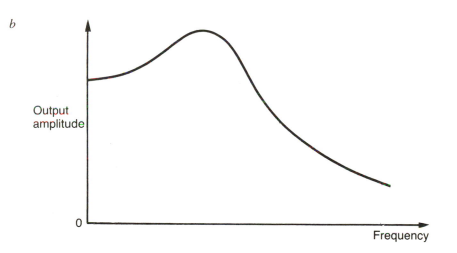

Fig. 4.8 *Responses of the servo-mechanism of Fig. 4.7 (a) to various inputs: (a) a slowly increasing angular displacement (curves (i)), and a single sudden turn (curves (ii)); (b) changes in the amplitude of output oscillations when the input is vibrated harmonically with a constant amplitude but different frequencies.*

in which petrol burned in a car engine could be better matched to the highly variable demands of car journey. Many other examples of *optimal control* arise in domestic, industrial and national contexts, where the problem is to make the best use of resources by ordered methods of control. To put the question in a general way, how should the supply of a costly resource, drawn from several sources having different price tags, be controlled so as to keep the overall cost as low as possible in the face of a changing demand?

One of the largest systems of this kind in the world controls the electricity supply to England and Wales. The figures involved are spectacular. At current values, the cost of fuel supplied to the power stations is about 4000 million pounds annually; altogether about 100 power stations containing a total of about 350 turbine generator sets are in play at any one time, of which about 50 are used variably to match supply to demand. When one realizes that the cost of

Fig. 4.9 *Summer and winter demands on the CEGB system in 1981–2 including days of maximum and minimum demand. The effects of load management on the peaks of demand are shown. Courtesy of the Central Electricity Generating Board.*

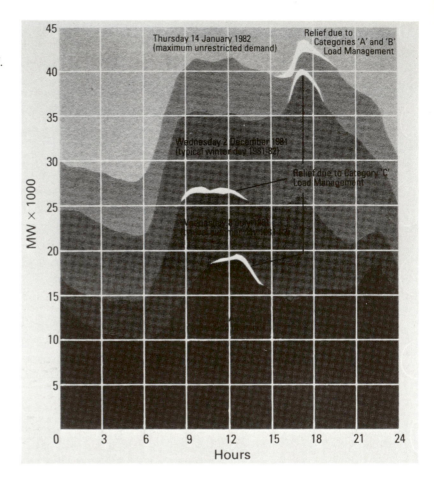

producing electricity differs significantly from one power-station to another and that there are large and rapid variations in demand depending on the time of day or night and on the weather (a change of 1°C changes demand by 600 MW, quite apart from the effect of wind and cloud), the heroic scale of the problem becomes apparent.

Let us look first at variations in demand. Fig. 4.9 shows how things fared over the 24 h of four typical days in the summer and winter of 1981/2. In each day a sharp rise between 8 a.m. and 9.a.m. as morning activities began was followed by a slight drop in the afternoon, and then by a second peak around 6 p.m. when meals were being cooked; overnight, demand returned to lower levels. The maximum daily variation was recorded in December, increasing from 21 000 MW in the night-time trough to 40 000 MW at the daytime peak. Over the year, from summer to winter, the *maximum* daily demand also changed by a factor of about two.

The effect of a special event, the royal wedding of 29th July, 1981, is shown in Fig. 4.10. In comparison with another national holiday in the summer of the previous year, the demand curve showed a slightly later start, but during the course of the television broadcast there was a pronounced drop as viewers on a massive scale switched everything off except their television sets. A succession of smaller peaks revealed

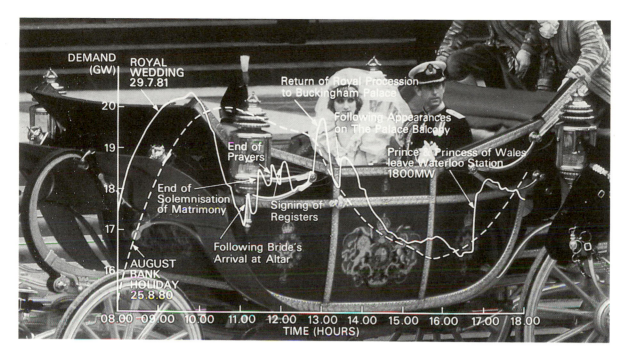

DEMAND (GW)

ROYAL WEDDING 29.7.81

Return of Royal Procession to Buckingham Palace

Following Appearances on The Palace Balcony

End of Prayers

Prince & Princess of Wales leave Waterloo Station 1800MW

End of Solemnisation of Matrimony

Signing of Registers

Following Bride's Arrival at Altar

AUGUST BANK HOLIDAY 25.8.80

08.00 09.00 10.00 11.00 12.00 13.00 14.00 15.00 16.00 17.00 18.00
TIME (HOURS)

how well synchronized the national demand was to particular incidents during the ceremony, and one might speculate that later peaks were caused by viewers catching up on daily requirements.

The national control room of the Central Electricity Generating Board is shown in Fig. 4.11. On the basis of predicted weather conditions and the availability of area supplies, the controllers use their experience to decide each day how much electricity should be generated in each of the areas into which England and Wales is divided, and they remain in continuous communication with the area controllers as the day goes on. They take into account the 'merit order' of the areas' power stations, measured in terms of cost per unit

Fig. 4.10 *Effect of screening the royal wedding, 29th July, 1981, on system demand. Courtesy of the CEGB.*

Fig. 4.11 *The control room for the supply of electricity to England and Wales. Courtesy of the CEGB.*

Fig. 4.12 *Transfer of electricity supplies (units of megawatts) between areas at a particular time, and the mains and a crystal clock. More often, the flow is from north to south. Courtesy of the CEGB.*

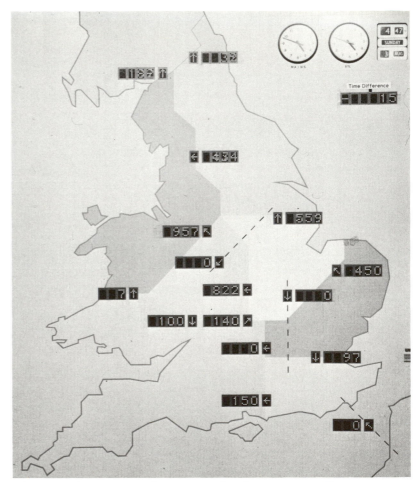

of electrical energy*, and set targets for the area controllers so as to minimize cost overall. It is then the task of the area controllers to call up the power stations, and the task of the power stations to respond by putting in or taking out particular generators. This in turn requires detailed control of such quantities as boiler steam pressures and turbine loading; as the speed of each machine must by law be kept within 1% of the nominal 3000 rev/min (in practice, much better control is usually achieved) and as the phase of each machine must remain in step with all the others, one can appreciate the importance and the scale of the problem. Fig. 4.12 shows the transfer of electricity between areas at another time: two clocks are also shown, one showing an apparent time as determined by the mains frequency, and the other a more accurate time based on a standard crystal frequency. The maximum allowable error of the mains clock is 15 s, but usually it is much less.

* The cost depends on many factors, including the fuel used. In 1983/ 4, the balance of UK supplies was 72% coal, 16% nuclear and 6% oil, the remainder including hydroelectric.

Methods of optimal control are also being developed to improve the fuel economy of petrol engines, which despite long experience are still pretty variable machines. Their requirements are complicated. In a conventional engine cylinder, maximum combustion efficiency is usually achieved at an air/fuel mass ratio of about 13/1 whilst noxious gases in the exhaust are minimized at a ratio of about 18/1. In the actual operation of an engine fed by a carburettor, there can be variations in these ratios from cylinder to cylinder and even in the same cylinder from cycle to cycle. And the precise instant when the spark should be delivered in order to burn the petrol–air mixture at maximum efficiency varies significantly with running conditions. When one remembers that the engine is subjected to cold starts, hot runs, rapid accelerations, sudden loads, periods of idling and so on, one can appreciate that optimal control is not a simple matter.

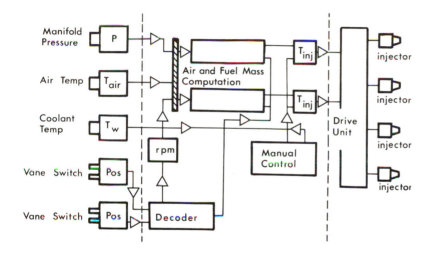

Fig. 4.13 *Lay-out of a control system for a four-cylinder petrol engine: a microprocessor supplied with information from the sensors on the left determines the optimum air/fuel ratio and timing for injection of the mixture into the cylinder. Courtesy of Ford, England.*

Fig. 4.13 illustrates the lay-out of a control system being developed for a four-cylinder car engine. An array of sensors measures the instantaneous values of the principal factors affecting performance and delivers corresponding electrical signals to a microprocessor. Armed with a predetermined criterion for the balance between efficiency on the one hand and low emissions on the other, the microprocessor carries out a rapid calculation of the amount of petrol required for peak performance in the prevailing conditions. Its output is transmitted to the drive unit, which in turn makes the appropriate delivery at the correct instant to an injector in each cylinder. If all works well, the engine will adapt itself continuously to achieve its full potential. Many difficulties have to be overcome in practice, not least the hostile environment created by the engine itself, but the results are promising.

Controls for making things

Applications of control to manufacturing continue to make dramatic advances. Ever since early engineers learned how to make useful products, methods of production engineering have been expanded and improved until now the subject is of enormous extent and importance. Never has its rate of change been more rapid than at present, and nothing has contributed more to its progress than improved controls.

Many different kinds of controls exist in a manufacturing company. Some, the managerial kind, are concerned with the supply and flow of materials, the organization of the work-force and the cost of operations; others are technical and deal with physical processes. All must work together harmoniously if the results are to be successful, and useful theories of control have been developed for complex systems of this kind, taking into account human as well as technical factors.

a

STEPHENSON'S LATHE

Fig. 4.14 *(a) George Stephenson's woodworking lathe of 1812 and (b) a modern milling machine.*

b

The work-horses of many factories are their machine tools. They come in all shapes and sizes, from bench-top drills to giant planing machines. Fig. 4.14 illustrates some of the changes that have taken place over the years. George Stephenson's woodworking lathe of 1812 was used to produce parts for his locomotives and was driven by a belt running over a large hand-driven pulley; a hand-held cutting tool took successive skims off the rotating work-piece until its diameter was reduced to the required size. A modern milling machine operates by applying a rapidly rotating hardened steel tool having many sharp cutting edges to the surface of the work-piece. In an edge-cutter, the cutting edges are arranged like teeth or flutes spiralled around the cylindrical surface of the tool; if a work-piece in the form of a flat horizontal plate of metal is moved past the fixed vertical axis of the cutter, the cutter will remove metal from the plate and leave behind a particular shape corresponding to the movements of the table. This is a versatile way of machining complicated parts to order, provided the work-piece can be moved as required. For cutting two-dimensional shapes, the work-piece is firmly attached to a table that can be driven horizontally in either or both of two perpendicular directions, say Ox and Oy. If suitably controlled, the table will follow any required path in the plane of motion and the finished work-piece will correspond to the user's intentions. Similar results can be achieved in three dimensions if the tool can also move vertically: Fig. 4.15 shows a general view of such a machining centre together with a fan-blade machined from a solid block by this means.

a

Fig. 4.15 *(a) A numerically controlled machining centre and (b) successive stages in the machining of a fan-blade.*

b

Evidently, much depends on the control of the work-table. Production engineers have borrowed a leaf from the automation of bygone days by programming its movements before machining begins so that the machine tool can carry out the whole process automatically. The program takes the form of symbolic instructions printed on punched or magnetic tape, or on magnetic discs, carefully prepared to generate signals for transmission to the drive units of the machine. In view of the nature of the signals, the whole process is known as numerical control, and if, as in the most advanced machines, a computer is included to work out the program, variations in the requirements from batch to batch can readily be introduced. Moreover, other important features of the cutting process such as depth of cut, speed of the cutting tool and rate of progress of the work-piece past the tool, all of which determine the quality of the result, can be incorporated in the control system.

Even more comprehensive results can be achieved by extending numerical control to machine tools designed to perform not merely one kind of operation but a continuous and elaborate sequence of operations, as shown in Fig. 4.16. As a complicated product such as an engine block requires drilling, boring, planing, grinding and other operations, the benefits of an automatic sequence become very attractive. Production engineers have developed machining centres that do just this. They are equipped with a store of cutting tools that can be selected automatically in an ordered sequence, fitted automatically in the drive unit and then applied to the work-piece in perfect synchronization with the controlled movements of the work-

Fig. 4.16 *A multi-tool numerically controlled machining centre: a succession of cutting operations can be performed on a workpiece by automatic selection of the appropriate tool.*

table. As much of the time spent in a manually controlled cycle is taken up in tool changing and in setting successive cutting conditions, the gains in speed and reproducibility can be dramatic.

Similar advances have been made in design and drawing offices, where traditionally much time has been taken up in the laborious task of producing detailed drawings for the workshops. Fig. 4.17 shows a small draughting machine in which the movement of the pen over the paper and the sequence of lines to be drawn are controlled by signals provided from a magnetic disc, pens of different colours being available for automatic insertion in the pen-holder. Once the required program has been recorded on the disc, an immaculate engineering drawing can be completed by simply depressing the start key. Moreover, variations in design requirements can readily be accommodated by making changes in the original program, and all variants can be retained indefinitely on discs for future use.

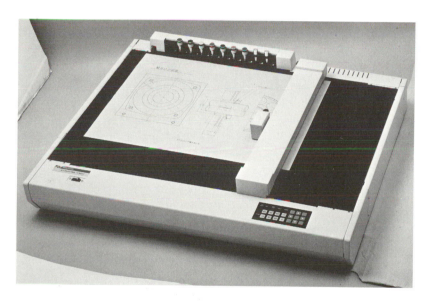

Fig. 4.17 *A small draughting machine capable of completing engineering drawings automatically when driven from a program on a magnetic disc. Courtesy of Environmental Equipments (Northern).*

Given that production drawings for the workshops can be drawn from programs on tape or discs, and that numerically controlled machine tools can make the parts from similar tapes or discs, the obvious question is whether the production drawings are necessary at all – could the program developed in the design office be presented directly to the machine tool without going to the trouble of making drawings? The answer is yes: direct transmission of programs is possible, and it will undoubtedly increase in importance in the future. But it must be remembered that machining operations are expensive, especially if they go wrong because of some fault in the program; no company wishes to speed the way to disaster by producing rejects, however efficient the process may be. Visual inspection of design information by an experienced production engineer will therefore still form a vital check on the proceedings, but he or she will depend increasingly on computer graphics for the purpose.

Robots

Robots have become one of the liveliest developments of our times. The term was first coined in 1923 by a Czech writer, Karel Capek, who introduced the word '*robotnik*', or worker, to mean a human-like machine. He followed a long tradition of writers who, with the sole aim of curdling the blood of their readers, had written stories about strange beings like Frankenstein's monster, superhuman in strength but lacking in heart and soul. That these ideas need not be taken too seriously is neatly illustrated by the Russian cartoon reproduced in Fig. 4.18.

Fig. 4.18 *A Russian impression of a young robot under instruction from* Let's Talk about Robots, *by I.I. Artobolevski and A.E. Kobrinski. Courtesy of The Young Guard, Moscow, 1977.*

For some 25 years or so, engineers have been busy developing real robots for work in industry, in space and the oceans, in general services and even in medicine. There is nothing even faintly monstrous about these machines. They are designed and made to carry out jobs that may be difficult, dangerous or just plain boring for human beings, just as innumerable other machines do. Viewed against the ranks of machines that operate more or less automatically, it becomes a little difficult to draw a hard-and-fast line between robots on the one hand and non-robots on the other – 'inherently versatile programmable devices' is perhaps as valiant an attempt as any – but, as usual, a satisfactory definition will probably follow at a respectful distance behind current practice.

Practical robots come in many different shapes and sizes, as illustrated in Fig. 4.19. Some trundle on wheels and others walk. Many work from a fixed base, using a series of links connected together like a human arm and with a gripper or a special tool at the end. A related species, tele-manipulators, are not robots in the strict sense of the word, but operate under the remote control of a human being who works at a safe distance from a dangerous environment, such as in nuclear installations or in space. Many powered devices have been developed to assist amputees in handling or in locomotion, whilst composite machines enable seriously disabled patients to carry out manipulative tasks by moving a tiny joystick, or even via control from the mouth.

But the great majority of robots at work today are used in industry, where the applications are legion. They handle heavy parts (hot or cold) paint, weld at a spot or along a seam, assemble components, feed other machines, inspect, test, run automatic warehouses, and generally speed the production process – even to the point of helping to make other robots. Naturally, they are not likely to show much benefit unless the production team has put them into a well-planned manufacturing sequence. In a new motor-car manufacturing plant at Longbridge, for example, more than 100 man-years of planning were absorbed in the preparations, but the results enabled completed cars to be assembled and finished in a hundred variants at the rate of one a minute.

In all these applications, robots provide automatic manipulation or locomotion. Their internal functions can be summarized as (i) controlling, (ii) sensing and (iii) thinking. Most current robots operate to a predetermined program of instructions that enables them to perform a sequence of tasks in strict rotation, such as applying a number of spot-welds to particular points on a car body. By simple adjustments of the instructions, a variety of patterns can be achieved so as to accommodate changes in the required welds, thereby offering much versatility. The programs can be worked out beforehand so that the electric motors or hydraulic actuators driving the robot receive the appropriate signals at the correct instants. Alternatively, a human operator can take the hand of the robot once through the desired sequence in order to record its movements magnetically for all future runs, as shown in Fig. 4.19(*c*).

Many difficult problems had to be tackled and solved before these apparently simple procedures could be used successfully in practice. They include rapid and accurate positioning of the robot's hand, accommodating the inevitable flexibility of light links, overcoming rather indeterminate friction in the joints and providing adequate power and dexterity.

A problem of particular theoretical and practical interest is the control of coordinated movements at the robot's joints. When we move our own arms to pick something up, the joints at the shoulders, elbows, wrists and hands combine their separate movements in a marvellously coordinated way, but it is much more difficult for an inanimate robot. The question to be solved is: if the robot has to place the object of its attentions in a particular position and orientation in space, how should its various joints move? In kinematic terms, the problem is to transform one set of variables (the coordinates of its hand) into another set (the joint angles). The relationships usually involve lengthy trigonometric functions, and the requirement may be subject to constraints imposed by mechanical interference. The computational problem is so severe that even large and costly computers are fully stretched to solve it quickly enough for on-line control. Alternative methods based on *rates of change* of the joint

a

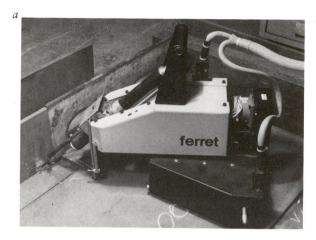

b

Fig. 4.19 Robots. (a) A wheeled
robot designed to weld ships'
hulls automatically at difficult
sites, including enclosed corners
(Newcastle University); (b) a
production line of robots in a car
factory; (c) a robot being taught
how to make welding runs
(Newcastle University);
(d) a stair-climbing robot
equipped with a spine-type arm;
(e) a research robot capable of
sensing the weight and the size of
the object to be gripped (Tokyo
University); (f) two robots
equipped with vision
collaborating in a delicate
operation; (g) powered arms for
patients with double-arm
disability (Medical Research
Council Powered Limbs Unit);

c

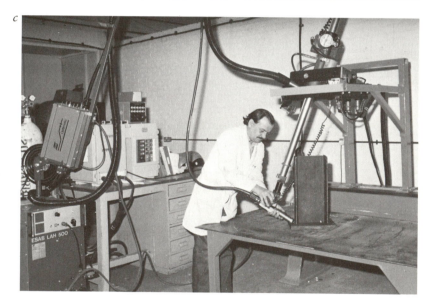

d

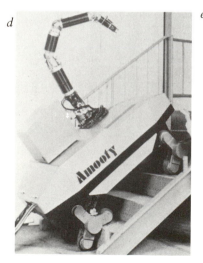

e

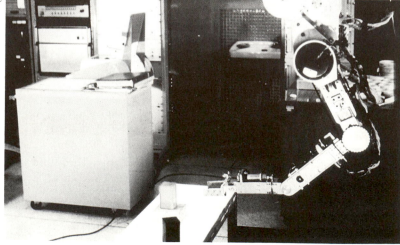

f

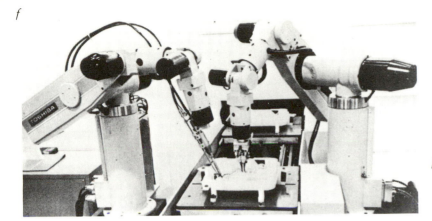

(h) *trials of a master–slave telemanipulator for work in a dangerous environment;* (i) *a general purpose machine for severely disabled patients (Heidelberg University).*

g

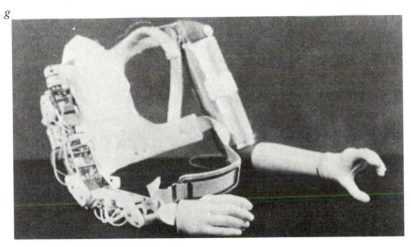

h

i

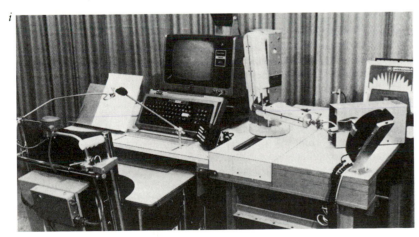

angles reduce the demand for computing considerably. These and similar procedures fall into the province of dynamics and control, where progress is continuing apace.

Even more striking advances are being made in the field of sensors. Human beings are equipped with sensors of astonishing capacities. They enable us to see, hear, feel, taste and smell the environment around us, and they keep us informed about balance, direction and the position of our limbs. Even a small fraction of this capacity would raise the performance of robots spectacularly, and researchers are seeing their ideas translated into successful achievement. Touch and forces sensors are now widely used, enabling the robot to detect obstacles, to gauge the dimensions of objects and to measure the forces it applies to the work-piece. Voice-recognition devices, capable of receiving oral instructions and turning them into command signals, are also being developed. But the most active area is vision, where the eyes of the robot are provided by cameras viewing the field and displaying the resulting picture on a television screen. To produce and to intepret the picture in a way that enables the robot to make its next move at an acceptable speed calls for high levels of physical imaging and software. We are at the beginning of advances that will shortly transform the scene.

The idea of thinking relates to the activities carried out in the robot's computer. It permits the robot to make decisions on the basis of built-in criteria, to use information gathered from its sensors, to plan the task ahead and to give command signals to the moving parts. Because extensive computation is required for even the simplest movements, a valuable premium is attached to fast procedures. A wide gulf separates preprogrammed robots and their successors – the so-called intelligent robots – capable of using on-line sensory information, but in view of progress in computers the future for intelligent robots is bright.

Fig. 4.20(*a*) shows an assembly of tubes in a development project concerned with the fabrication of boilers for a nuclear power station. In the power station, water pumped through the tubes is heated by hot gases passing over the tubes, and the steam so generated is used to drive the station's turbines. Fabrication of the assembly called for stringent control of the manufacturing process, including the welding of the spacers – small pieces of steel forming the transverse bands – to

Fig. 4.20 *Stages in the development of a vision-controlled welding system for the fabrication of boiler tubes: (a) an experimental rig for the positioning of the welding-head on the line of the seam; (b) part of the production facility; (c) the arrangement of the light source and the camera; and (d) the image of the weld-area on a TV screen. Courtesy of NEI Nuclear Systems and Newcastle University.*

a

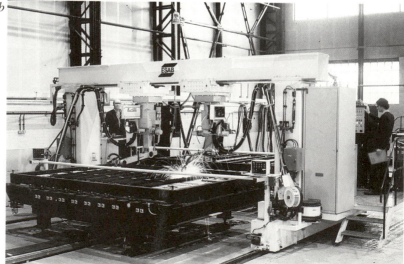

b

c

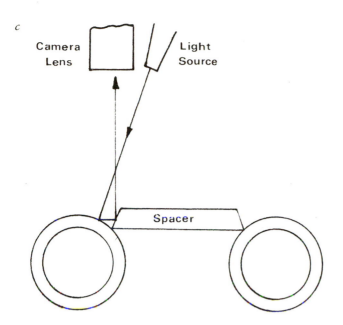

Camera
Lens

Light
Source

Spacer

d

the tubes. As a large number of assemblies and several million spacers
were involved, the company installed a robot-welding facility, part of
which is shown in Fig. 4.20(*b*). The robots were preprogrammed to
complete a weld and then to move to the next spacer, and to speed the
process a vision system was developed as shown schematically in
Fig. 4.20(*c*) and physically in Fig. 4.20(*a*). A ray of light reflected from
the bright machined edge of a spacer enabled a television camera to
display the picture shown in Fig. 4.20(*d*) on a screen, and a computer
then instructed the welding-heads of the robots to position themselves
accurately on the required line of the weld.

A PUMA robot (Programmable Universal Machine for Assembly),
Fig. 4.21, showed its paces. First, it demonstrated movements of
individual joints by single rotations – waist (vertical axis), shoulder
(horizontal axis), elbow (second horizontal axis), three rotations
about perpendicular axes at the wrist, and closure of the gripper.
Next, by means of coordinated movements of its joints, it moved the
gripper along oblique lines in space, and then, to complete the
preliminaries, it moved a pointer fitted in its gripper along a
prescribed path to touch a fixed marker; after withdrawing and
performing a rapid whirl in space, the pointer was once again speedily
returned along the same path to touch the marker neatly in a
convincing demonstration of accuracy and reproducibility.

The final demonstration took the form of a competition. A young
volunteer, Douglas, was invited to test his manipulative skills by
moving a stiff wire, bent into a small ring at its end, around another
stiff piece of wire of a peculiar shape. Every time the ring touched the
wire, an electrical circuit was closed and a loud buzzing noise could be
heard. Douglas's task was to move the ring from end to end of the
circuit with as few buzzes as possible. This he did with much skill, but

Fig. 4.21 *The PUMA robot and a young challenger. Courtesy of Unimation (Europe).*

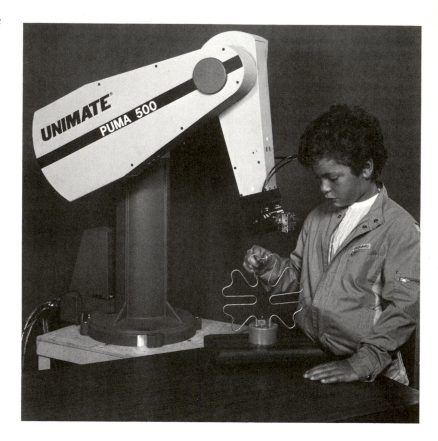

the task was so difficult that a few buzzes were inevitable. PUMA, programmed for the event, began badly with several buzzes on the first few turns. Then, under new instructions, it began in earnest. It not only raced around the circuit on a clear run but also stopped halfway to perform some exuberant manoeuvres with its links while holding the ring clear of the wire. To round things off, it laid down the ring, switched off the electrical circuit, and bowed gracefully to the audience.

5

FLUIDS AND FLIGHT

Working fluids

If all the fluids in the world were suddenly frozen solid, everything would stop; no ship, aircraft or vehicle could move, engines would shudder to a halt, sources of power and heat would fail, life forms would perish and the Earth itself would abandon its ordered path. Ever since Archimedes leaped out of his bath 2000 years ago, the behaviour of liquids and gases has continued to absorb the attention of scientists and engineers; although his famous cry 'Eureka!' opened up an understanding of the laws governing the *static* characteristics of fluids, vital physical links accounting for the *dynamical* action of fluids were discovered only within living memory. As we shall see, many such discoveries followed the invention and development of new machines rather than the other way around.

An excellent example of something driven by the static force of a fluid is the hot-air balloon. This early venture to take to the air was marked by the successful flight of the Montgolfier brothers in Paris some 200 years ago. A somewhat shorter flight was launched in the lecture theatre as shown in Fig. 5.1(a). When the air inside the canopy was heated by a burner, the balloon rose majestically to the ceiling, carrying with it a small passenger. The hot air in the canopy, being slightly less dense than the surrounding atmosphere, had provided enough buoyancy to overcome the weight. As the air cooled, it lost buoyancy, the balloon descended, and the passenger returned unharmed to the ground.

Hot-air balloons go up. Parachutes come down. Their motion depends to a greater extent on the dynamics of the air flowing past the canopy, as anyone will realize who has watched the skilful drops of trained parachutists controlling their descents. They contrive to make their motion through the air generate both lift and sideways thrust, although it must be admitted that the two model parachutists came down pretty straight when released from the ceiling of the lecture theatre, as shown in Fig. 5.1(b).

One of the earliest machines used for pumping fluids is the simple and effective pump known as the Archimedean snail and shown in Fig. 5.2. Its action is beautifully smooth and direct, and it is still used for purposes of irrigation in some parts of the world. When the handle of the model was turned, water was scooped up at the lower end of the open spiral tube; driven upwards by means of a clever combination of centrifugal action and the kinematics of the screw, it travelled along the tube to discharge at the top. From there, in the real thing, the water flowed down channels leading to the fields.

Another pump developed much later is also based on centrifugal action. It is used in many applications extending from small units in

Fig. 5.1 (a) *A hot-air balloon going up, and (b) parachutes coming down.*

a

b

Fig. 5.2 *A model of an Archimedean snail-pump: water is scooped up at the lower end and discharged at a higher level.*

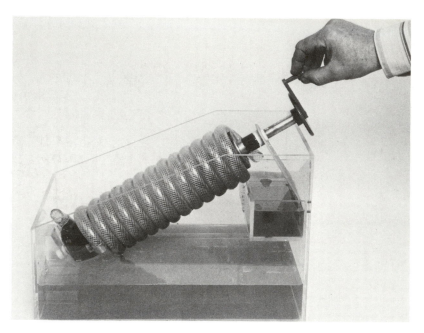

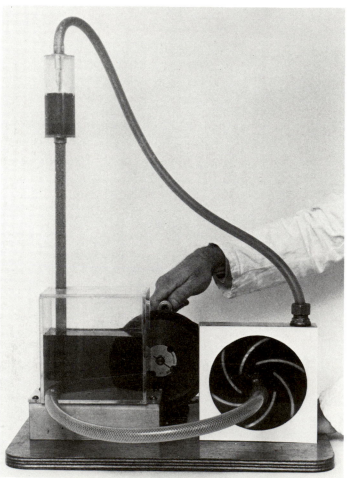

Fig. 5.3 *A simple centrifugal pump. The curved blades behind the glass plate are attached to a flat disc which rotates inside the rectangular body of the pump: water from the reservoir is drawn into its centre and forced upwards to discharge at the top.*

domestic central heating installations to giant machines supplying feed-water to the boilers of power stations. Fig. 5.3 illustrates a simple version.

The first machine that provided useful amounts of power was probably the water-wheel. As early as the Domesday Book of 1086, more than 5000 water-mills were running in the southern half of England. By the time of the Middle Ages, almost every village that had a stream also had a water-wheel. The model in Fig. 5.4 shows an undershot wheel driven by the water flowing past the submerged lower paddles; in an overshot wheel, the water adds its weight to the propulsive torque as it flows over the top.

Windmills, too, have a long history. Fig. 5.5(*a*) shows a typical model. They not only ground corn but also supplied power for pumping, for sawing wood and for many other useful purposes. Many are still in use, but as more competitive sources of power became available they passed their hey-day at about the same time as the water-wheels. Now, with the help of modern technology, they are enjoying a revival as generators of electricity. Fig. 5.5(*b*) shows such a machine in Orkney; the blades measure 20 m from tip to tip and, when running at speeds of 80–100 rev/min, generate about 250 kW of electricity for the national grid. Features of the design include controls

Fig. 5.4 *A model of an undershot water-wheel.*

a

b

Fig. 5.5 *Windmills ancient and modern. (a) A model equipped with a fan-tail designed automatically to turn the cap on which the sails are mounted into the winds, and (b) an aero-generator in Orkney capable of generating 250 kW of electrical power.*

to position the blades head-on to the prevailing wind, over-speed controls, and microprocessors to match the electrical output of the generator to the requirements of the grid. Larger machines with electrical outputs measured in megawatts are also under design and construction, some employing blades rotating about a vertical rather than a horizontal axis. As engineers explore the prospects of extracting useful energy from the wind, many new and exciting possibilities are being opened up.

Power can be transmitted through fluids in many ways. The simplest is the hydraulic press, originally patented by the engineer Joseph Bramah in 1796 and illustrated in Fig. 5.6. When the small piston is pushed downwards, the pressure generated in the cylinder is transmitted to the larger cylinder through the connecting pipe; the same pressure creates a much larger force on the large piston, capable of raising a heavy weight or, as in Bramah's application, forging metal parts. The same principle has been used successfully to transmit power over long distances. In London and some other large cities, pumping stations supplied pressures of up to 70 atmospheres to networks of subterranean water-pipes; the power delivered at the receiving ends drove a variety of hydraulic engines that raised lifts or worked dockside machinery. At peak usage, about 200 miles of pipe were in operation below London, but, although the pipes are still there, they made their last delivery a few years ago.

In the Bramah press, the small piston has to travel further than its large companion because the transferred water occupies a shorter length of the large cylinder. If, on the other hand, we imagine that the *large* piston is pushed, water would flow rapidly through the connecting pipe, and if the small cylinder were removed entirely the water would issue from the pipe as a jet – the smaller the pipe, the faster the jet. Any aperture through which pressurized fluid can escape will provide a jet, sometimes, as in firemen's hoses, on an impressive scale, more modestly in garden hoses – and also in water-pistols. After inspecting the small hole in a loaded water-pistol, a cheerful volunteer – who in the interest of science had agreed to take part – remained smiling as a jet of water gave him a brief but accurate shower from a distance of several metres, Fig. 5.7.

Jets of fluid also drive machines. When a jet of water strikes a stationary flat plate at right angles, the forward motion of the water is destroyed and it flows over the surface to the edges: the force the jet exerts on the plate is proportional to the rate of change of the water's forward momentum. A larger force can be generated by replacing the flat plate by a hemispherical cup facing the jet; as the water strikes the centre of the inside of the cup, it is turned back on itself and streams off the edges with much the same linear momentum as during the approach, but now in the opposite direction. The rate of change of its linear momentum is double that caused by the flat plate, and the cup is subjected to twice the force.

Two experiments showed how things worked. Both were based on the pressure supplied by the water-mains to generate jets through identical nozzles, Fig. 5.8 In the first, the jet struck a circular flat plate placed in a horizontal position at a small distance above the nozzle. When the jet was fully developed, the water streamed horizontally off the edges of the plate, demonstrating convincingly that all its upwards linear momentum had been lost. Moreover, the plate was attached to a horizontal lever equipped with a sliding weight, like an old-fashioned weighing machine, so that the upwards thrust of the jet could be measured by moving the weight along the lever until horizontal balance was restored. In the second version, a hemispherical cup replaced the flat plate; the jet was turned back on itself and poured downwards off the edge of the cup in an attractive

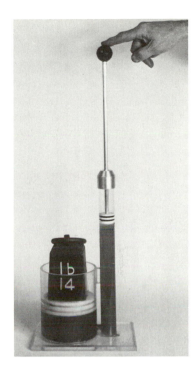

Fig. 5.6 *A version of Bramah's hydraulic press.*

Fig. 5.7 *An application of a water-jet.*

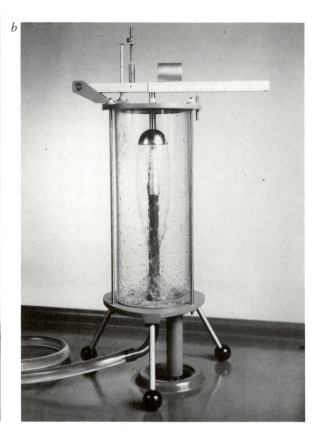

Fig. 5.8 *A water-jet supporting loaded levers. In (a) the jet struck a flat plate and in (b) a hemispherical cup: because the rate of change of the jet's linear momentum was twice as large in (b), the balance weight on top of the lever had to be moved twice as far to keep the lever horizontal. Courtesy of Techquipment.*

Fig. 5.9 *A model Pelton wheel driven by a water-jet. Courtesy of Techquipment.*

symmetrical flow. It was found that the sliding weight had to be moved along the lever about twice as far to restore horizontal balance, confirming that the force on the cup was about double that on the plate.

Water jets impinging on curved cups or buckets form the basis of large machines known as Pelton wheels, which are used where a high head of water is available, as in the fall from a reservoir, to drive electrical generators. Their specially shaped cups, invented in the Calfornian gold-fields, incorporate a central partition which improves performance significantly by dividing the jet into two streams and providing a smooth transition from forward to backward flow. Naturally, the motion of the cups affects the exchange of forward and backward momentum of the water, but it is easy to show mathematically that, in order to generate the largest achievable driving torque, the speed of the retreating cup should be half that of the advancing jet.

The model Pelton wheel shown in Fig. 5.9 did not have the benefit of a high head of water but instead was driven from an ordinary water tap. Nevertheless, it accelerated so rapidly that it quickly became a blur, leaving behind only a high-pitched whine.

Flying machines

The most striking application of fluid dynamics is to aeronautics, the foundations of which were laid by the astonishing studies of Sir George Cayley around 1810. They were astonishing for several reasons. First, they were almost entirely single-handed; second, he concentrated on fixed-wing machines as opposed to the popular but totally unsuccessful attempts to copy the flight of birds by using flapping wings; third, he identified the important and separate ideas of lift, propulsion and control (he proposed rear elevators, a fixed vertical tail and rudder, and airscrews), and lastly he conducted meticulous experiments on each of the chief elements. He was an expert mathematician and prolific inventor but he was so far ahead of his time that practical applications took some time to catch up – no less than three generations in the case of powered and manned flight.

A full-scale copy of an experimental rig used by Cayley to measure the lift on surfaces moving through air was demonstrated in the lecture theatre, Fig. 5.10. It consisted of a horizontally whirling arm of wood, balanced like a see-saw on a hinge at the top of the vertical drive-shaft, and carrying a light rectangular plate of stiffened paper at one end. The plate could be rotated about the long axis of the arm and fixed there at any chosen inclination. When the arm was driven around (by a falling weight in Cayley's experiments, manually in the demonstration), the air exerted forces on the plate dependent on the inclination of the plate to the direction of its motion. The purpose of the original experiments was to measure these forces, in particular the lift force tending to tilt the arm upwards. At first, the plate was fixed in a horizontal position on the arm, so that it sliced through the air parallel to its own motion: it was found that the arm continued to

Fig. 5.10 *Demonstration of a full-scale replica of the whirling-arm apparatus used by Sir George Cayley. Courtesy of the Science Museum.*

whirl in a horizontal plane without any apparent lift force arising from air pressure. Next, the plate was tilted slightly about the long axis of the arm so that its front or leading edge was up and its back or trailing edge was down. When the arm was again whirled around, the arm rose upwards as it experienced the lift force of the air. By such means Cayley was able to establish accurate values of aerodynamic forces, and to relate them to the inclinations of the moving bodies, their shapes and their airspeeds.

On 17th December, 1903, a century after Cayley's experiments, the most famous event in the history of aeronautics took place near Kitty Hawk in North Carolina. Orville Wright, who had won the toss from his brother Wilbur, took to the air in a powered heavier-than-air machine for 12 s and flew 120 ft. Within a year, the brothers were flying distances of over 20 miles, to the consternation of their competitors in other parts of the world. Their achievements were no sudden breakthrough. They were the result of years of dedicated work involving hundreds of glider flights, wind-tunnel tests, engine and airscrew developments and, perhaps most original of all, studies of control methods. The Wrights had realized, more clearly than anyone else, that the way ahead was to make the combination of man and machine controllable and stable in flight – the machine did not have to be stable by itself. Perhaps they had been influenced by their experience of making bicycles in Dayton, Ohio. A few years later, a comprehensive theoretical treatment was published by Frederick Lanchester which remained the definitive work on the subject for many years.

Fig. 5.11(*a*) shows the historic first flight, with Orville as pilot lying on the lower wing, and Fig. 5.11(*b*) a model of the aircraft. The elevators were mounted at the front, the rudder at the back, and two contra-rotating pusher propellers, driven by chains from a four-cylinder in-line petrol engine, were placed behind the wings. Of special interest was the way in which the construction provided lateral

Fig. 5.11 (a) *The first manned and powered heavier-than-air flight at Kitty Hawk, North Carolina, on 17th December, 1903; (b) a model of the aircraft looking from the tail.*

control: by means of cables controlled by the pilot, the outer surfaces of the wings could be warped to generate aerodynamic forces upwards on one side and downwards on the other, an idea that anticipated the ailerons of later design.

Advances in flying in the years that followed were dramatic. Blériot flew across the English Channel in 1909, the 1914–18 war accelerated development even faster, and in 1919 Alcock and Brown achieved the first flight across the Atlantic. The new technology of wind-tunnels was developed in this period to study the complicated flows of air over surfaces, at first with models and later with full-scale aircraft. As a result of this work, many entirely new discoveries were made which soon led beyond the limits of subsonic flight into the as yet unexplored régime of supersonic aerodynamics.

The flow of fluids

Anyone who has watched a large, heavy aircraft take off or land must occasionally have marvelled at the way it is held up in flight by nothing more substantial than thin air. If that thought should also occur occasionally to passengers, only to be thrust rapidly to the back of their minds, they might do worse than reflect that they are kept safely aloft by a fundamental property of fluid flows, which roughly speaking can be expressed as 'the faster the flow, the lower the pressure'. To see how wings keep aircraft up, let us consider first a simple experiment involving coloured water flowing through a tube, as illustrated in Fig. 5.12.

The tube, known as a venturi meter and used in practice to measure the rate of fluid flows, was constructed with a short section of smaller

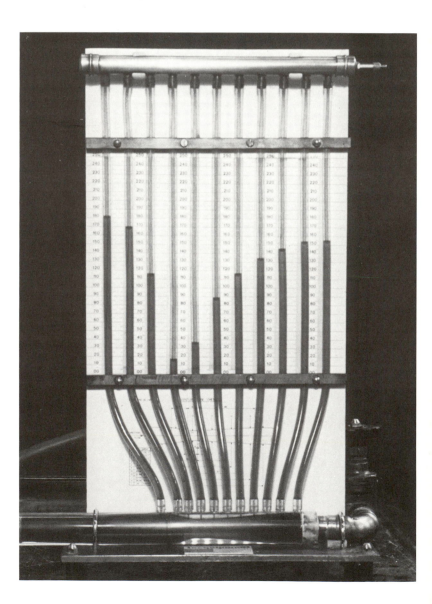

Fig. 5.12 A venturi meter for measuring the rate of fluid flows. The height of the water in the vertical tubes shows that the pressure in the neck, where the flow is faster, is lower than the pressure in the full section: the difference provides a measure of the rate of flow.

diameter than the rest, so that the water flowed faster through the neck than elsewhere. It was equipped with a number of thin vertical tubes called manometers connected to small holes in the main horizontal tube and open to the atmosphere at their upper ends. As the tap controlling the inlet was opened, the water flowed through the tube to the outlet, and soon the manometers began to fill with water. When the flow was fully developed, it was seen that the columns of water in the manometers remained stationary at different heights, those connected to the neck being markedly lower than the others. This revealed that the pressures supporting the water columns were lower at the neck, where the stream velocity was higher, than elsewhere. The relationship is of great importance in fluid dynamics, and for flows involving incompressible and frictionless fluids (the effects of compressibility on water are significant only at very high pressures) it can be expressed mathematically as

$$\frac{1}{2}\,\rho u^2 + p + \rho gh = \text{constant}$$

where ρ is the density, u the velocity, p the pressure, and h the height above a chosen datum. The equation was derived by Daniel Bernoulli in 1738 on the assumption that the sum of the kinetic and potential energies of the fluid remained constant, all forces except normal pressures and gravity being excluded from the account. It has since been extended considerably to account for more elaborate flows, but it remains a cornerstone of fluid dynamics.

Let us apply this idea to the flow of air over a wing section or aerofoil, Fig. 5.13, making the reasonable assumption that for moderate air speeds, up to a few hundred miles per hour, the compressibility of the air may be neglected. It is convenient to imagine the wing fixed and the air flowing past it, as in a wind-tunnel, rather than the other way around; in both cases, the forces on the wing will be the same. Two other simplifications complete the picture: we take the flow to be horizontal, thereby eliminating the minor effect of gravity represented by the last term in Bernoulli's equation, and we exclude for the moment any effects arising from friction of the air. Naturally, the airflow must divide as it encounters the obstacle in its path, and if the wing is well designed its shape will ensure that the air flowing over the upper surface travels faster than the air flowing over the bottom surface. According to Bernoulli, this means that the air pressure is less on the upper surface than on the lower, and the result

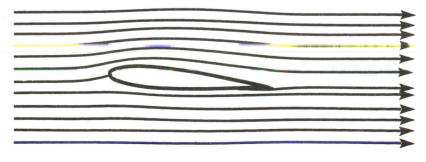

Fig. 5.13 *Streamlines of subsonic flow past an aerofoil.*

is an upward force that supports the weight of the aircraft. The precise shape of the aerofoil is naturally of great importance, and aerodynamicists have developed advanced techniques to match the shape to design specifications.

Many more complicated factors are encountered in practice than our simple picture suggests, most of which are associated in one way or another with the internal friction or viscosity of the air. Its most direct effect is to exert drag forces on the surfaces over which it passes, but as we shall see it can also affect the entire flow pattern around and behind the aerofoil, thus contributing additional and often considerably larger drag effects. Overall, a leading measure of performance is the ratio of lift force to drag force, which for modern subsonic aerofoils in cruising conditions range from about 20 for large aircraft to over 60 for gliders. Limited but useful amounts of lift can be generated simply by inclining a flat plate to a flow of air, as was illustrated by a young volunteer who catapulted a model glider with flat wings towards the audience, Fig. 5.14. It was followed by a simple version of a helicopter rotor, which developed such a good lift when its rotor was spun at take-off that it crashed into the wall of the theatre some 20 ft up.

For a chosen section, both lift and drag depend on the angle the aerofoil makes with the direction of flow – the angle of attack – as demonstrated by means of a smoke-tunnel, Fig. 5.15. It consisted of a sheet-metal chimney up which smoke flowed from entry at the bottom to exit at the top. A thin tube projecting horizontally into the airstream near the bottom released several spaced-out filaments of paraffin smoke which marked the details of the flow past a model aerofoil mounted behind a window in the working section. With the rate of flow set at a particular value, the angle of attack of the aerofoil was gradually changed by hand from zero (head-on) to progressively larger values, and the consequent changes in flow patterns were observed through the window. At first, the filaments of smoke divided

Fig. 5.14 *Launch of an unmanned unpowered aircraft.*

smoothly as they met the aerofoil, Fig. 5.15(*a*), following streamlines that remained close to the solid surfaces. But, as the angle of attack increased, the flow became disturbed around the trailing surface on the low-pressure side. As the angle was increased still further, there came a point at which the flow in that area changed dramatically – the streamlines broke away completely from the surface and left a confused eddying flow in command over a substantial part of the surface, Fig. 5.15(*b*). In such circumstances, the pressure distribution on the surface bears little resemblance to that corresponding to smooth streamline flow, and from a practical point of view the most important difference is that the lift collapses – the aerofoil stalls. A classical picture of this phenomenon is shown in Fig. 5.15(*c*).

Fig. 5.15 *The smoke-tunnel experiment. Filaments of smoke released into a vertical airstream marked the pattern of flow around a stationary aerofoil. In (a) the air flowed smoothly over the surfaces, but at a high angle of attack, as in (b), the flow separated from the low-pressure (right-hand) side in a confused eddying manner. A classical picture of the phenomenon is shown in (c), from Prandtl, 1930.*

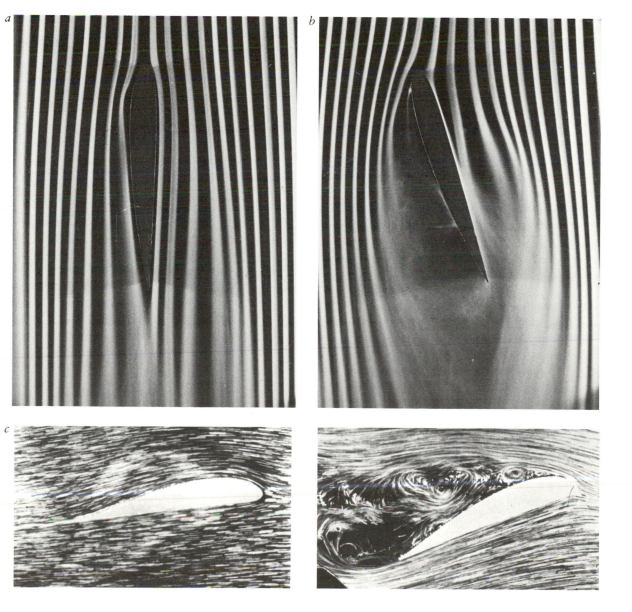

When we consider the details of this transformation, we enter one of the most difficult and interesting areas of aerodynamics. The simplifying assumptions leading to the form of Bernoulli's law expressed earlier must be modified to take account of *viscosity*, a property that exists to a greater or lesser extent in all real fluids. It can have dramatic effects on fluid flow.

Some liquids such as glycerine and treacle are very viscous at ordinary temperatures; others such as water and petrol are much less so, and gases have yet lower viscosities. In more precise terms, viscosity expresses the forces that resist the sliding or shearing of fluid elements over each other. If we imagine an element of fluid distorting like a pack of cards when each card slides over the next, the viscous force between successive layers is found to be directly proportional to the coefficient of viscosity of the fluid (usually simply called viscosity), to the areas of the sliding layers, and to the sliding velocity per unit distance across the thickness of the pack. Unlike solids, fluids have no resistance whatever to pressures applied perpendicularly to the surfaces of their elements, but viscosity provides them with a resistance to shear, provided a shearing motion actually occurs: if the motion stops, so does the resistance – the forces at work behave more like friction than springs. For their sources, we have to look at what happens to the molecules of fluid; for gases, molecules jumping between adjacent layers in relative motion are chiefly responsible, whereas in fluids, where the molecules are more densely packed, inter-molecular forces between the layers also play a significant part.

Viscosity is important in many kinds of fluid flow as well as in aerodynamics, and we digress briefly to consider a remarkable and important discovery concerned with the flow of liquids in pipes. If one watches a stream of water issuing from a tap, its appearance may be seen to change from transparent smooth flow to a milky turbulent flow as the tap is opened. Similar changes occur in flows in pipes. However, the conditions governing the transition remained a mystery until, in a series of famous experiments carried out in 1883, Osborne Reynolds discovered that the transition occurred when a certain non-dimensional parameter, now known as the Reynolds number, reached a particular value. That number, one of the most important numbers in fluid dynamics, is defined for pipe flow as

$$(Re) = \frac{vd\rho}{\mu}$$

where v is the stream velocity, d is the diameter of the pipe and ρ and μ are the density and viscosity of the fluid, respectively. Provided (Re) is less than 2000, the flow remains streamlined, whereas for higher values of (Re) it becomes turbulent. Similar conditions apply to open flows over geometrically similar aerofoils, where the Reynolds number is defined in terms of a characteristic length of the surface. Several important conclusions follow at once: turbulence is associated with higher velocities of flow, with larger dimensions in geometrically similar systems, and with fluids having greater densities and smaller viscosities. The roughness of the surface also plays a part in the proceedings, but in pipe flows its chief effect is to increase the frictional force on the solid surface *after* the onset of turbulent flow, rather than to change the point of transition – although, as we shall see, there are other curious effects of roughness in certain applications.

A version of Reynold's experiment was demonstrated with a vertical glass tube through which water could flow at a controlled rate from a tank full of water at its head, Fig. 5.16. A thin filament of dye released into the water at the top marked the progress of the flow through the tube. At slow rates, the filament kept to a straight and

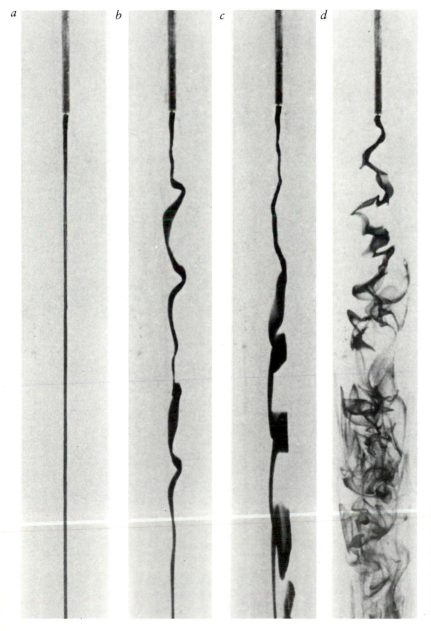

Fig. 5.16 *Development of turbulence as the rate of flow of water down a vertical pipe is increased. The line of dye released from a thin tube at the top of the picture remained smooth and straight at low rates (a), but became progressively more turbulent as the rate was increased.*

narrow path down the centre of the tube but as the rate of flow increased it became unstable, flicking from side to side until, as the point of transition was passed, it broke up altogether into a confused, turbulent pattern. In these conditions, random movements of particles of the fluid occur between adjacent layers in such a complex way that mathematical analysis is possible only on a statistical basis, and then only to a limited extent. Instead, the *average* characteristics of the flow have to be determined experimentally for use in any ensuing calculations.

Viscous forces are particularly important in the regions of flow next to a solid surface. Even when the surface is extremely smooth, molecules trapped in its tiny irregularities ensure that the relative velocity of the stream at the surface is zero (no slip), and that shearing motion also occurs in successive outwards layers. This remarkable and universal phenomenon led Prandtl to enumerate the theory of *boundary layers* in 1904. It enabled flows near surfaces to be considered in two regions, the boundary layer extending outwards from the surface to the region of the mainstream flow, and the mainstream itself; in the boundary layer, shearing actions and their viscous forces predominate, whereas in the mainstream they are negligible. Even when part of the boundary layer is turbulent, there is still an extremely thin layer next to the surface, the so-called laminar sublayer only a few micrometres thick, where turbulence is suppressed and smooth or laminar flow prevails. The discovery of these phenomena marked one of the most striking advances in fluid dynamics.

The actual thickness of boundary layers varies widely with the viscosity of the fluids involved; it may be as little as a few millimetres or less for lightly viscous gases such as air and as much as several centimetres for heavily viscous liquids such as glycerine. Fig. 5.17 shows a plastic tank containing glycerine flowing out through a raised gate at one end. Before the flow began, a syringe was used to lay down thin lines of dye in the glycerine to mark the progress of the flow, one across the tank near the surface, and another vertically from bottom

Fig. 5.17 *Highly viscous flow. As the glycerine flowed out of the tank beneath the raised gate, the lines of dye marked its progress: in (a) the horizontal line showed the pattern of flow at the surface, and in (b) the vertical line, placed at the centre of the tank, showed how the flow varied with depth.*

a1
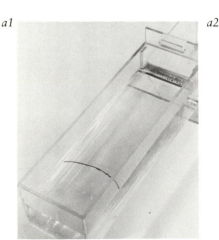

a2
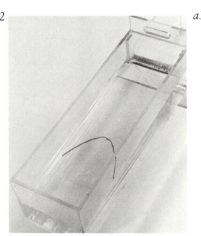

a3
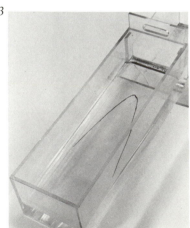

b1

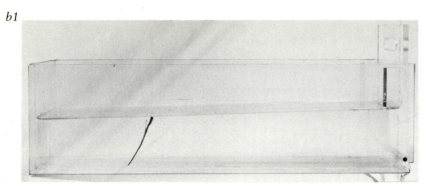

b2

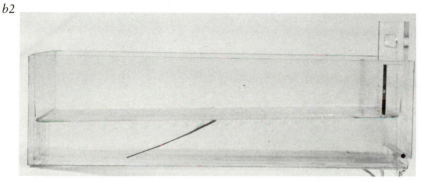

b3

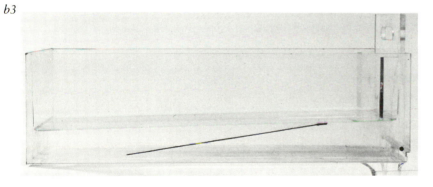

to top. When the gate was opened to release the flow, the lines of dye, moving with the glycerine, revealed that across the tank the fluid remained stationary at the sides while the central regions slid past. A similar result was observed with the vertical dye line, which remained attached to the bottom of the tank but bent progressively in the direction of flow as the upper layers slid forwards faster. As a reference to the Reynolds number of the flow (low velocities, high viscosity) would have predicted, no turbulence was visible anywhere.

With these ideas in mind, we return to our stalling aerofoil. At first, a closely attached boundary layer enveloped the entire surface and smooth flow prevailed. As the angle of attack increased, the mainstream flow over the low-pressure side moved progressively faster, thereby offering lower pressure and increased lift; but, as it combined with the slower flows over the other side, it was obliged to decelerate once again as it met the higher pressure at the trailing edge.

Meanwhile, the boundary layer, sandwiched between the solid surface and the mainstream flow, was finding increasing difficulty in advancing against the rapid increase in pressure at the rear, and eventually was forced back on itself, thereby separating from the aerofoil altogether in a confused eddying motion. Naturally, if the boundary layer had been more persistent in its advance against the adverse pressure gradient, separation and stall would have been delayed, and if, in particular, the boundary layer had been turbulent, the increased kinetic energy associated with such random motion would have enabled it to remain attached for longer as the angle of attack was increased. The precise conditions governing transition from laminar to turbulent flows in boundary layers depend primarily on the relevant Reynold's number, and are of great importance in establishing lift and drag forces on aerofoils. Moreover, various devices have been developed to delay boundary-layer separation, including slots placed at the leading edge to reinforce the boundary layer downstream by directing air from the lower to the upper surface, and porous surfaces which suck the boundary layer inwards.

The final aerodynamic experiment was based on a small wind-tunnel, Fig. 5.18, equipped with an aerofoil that could be rotated by hand so as to vary its angle of attack relative to the airstream. Thirty-six small holes drilled into its upper and lower surfaces were connected through thin plastic tubes to an equal number of manometers. Each registered the prevailing air pressure at the corresponding point of the aerofoil surface, the manometers on the left of the picture being connected to the upper surface and those on the right to the lower surface. Sucking and blowing through a temporarily disconnected tube confirmed that *higher* pressures corresponded to a *lowering* of the liquid levels in the manometers, and conversely that lower pressures corresponded to higher liquid levels. At first, the aerofoil was positioned head-on to the airstream; pressure variations over the top and the bottom surfaces were then seen to be

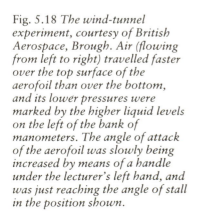

Fig. 5.18 *The wind-tunnel experiment, courtesy of British Aerospace, Brough. Air (flowing from left to right) travelled faster over the top surface of the aerofoil than over the bottom, and its lower pressures were marked by the higher liquid levels on the left of the bank of manometers. The angle of attack of the aerofoil was slowly being increased by means of a handle under the lecturer's left hand, and was just reaching the angle of stall in the position shown.*

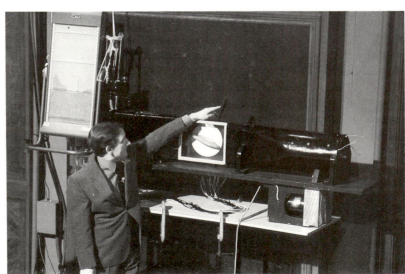

almost symmetrical, higher at the leading and trailing edges and lower
in between. As the angle of attack was gradually increased, things
changed markedly. For the first 20° or so, the pressures over the front
part of the upper surface dropped considerably compared with those
on the lower surface, showing that lift force was being generated on
the aerofoil as a whole. As the angle of attack was increased further, a
rapid change occurred – the manometer levels connected to the upper
surface near the leading edge suddenly collapsed. The boundary layer,
faced by an increasing pressure gradient as it moved along that
surface, had separated. If that had happened in flight, the aircraft
would have stalled.

Fluid gamesmanship

Skilled games players have learned through experience how to make
good use of aerodynamics. Consider the strange case of dimples in
golf-balls, where the boundary layer plays a critical if somewhat
paradoxical rôle in the proceedings. Early golf-balls were made
smooth, on the reasonable assumption that the smoother the ball, the
further the drive from the tee. Later on, exactly the opposite was
found to be true, and although no clear record exists of how it was
discovered – a dog addicted to chewing golf-balls may have played
some part in the proceedings – it soon became apparent that the ball
would go further if its surface were deliberately roughened. And so
dimples made their appearance. Fig. 5.19 illustrates the aerodynamics
involved.

At the speeds at which the golf-ball leaves the tee, the boundary
layer remains laminar around the front side of a smooth ball but
separates from the surface near the maximum width, thereby creating
a large, turbulent and low-pressure wake. Because the total drag on
the ball depends not only on the friction of the air but also on the
pressure difference between the front and the back, things would be
much improved if the boundary layer could be persuaded to envelop
more of the back surface and so leave behind a smaller wake. This is
the purpose of the dimples. By deliberately disturbing the boundary
layer on the front side, they create turbulence in the boundary layer

Fig. 5.19 *The case for dimples in
golf-balls.*

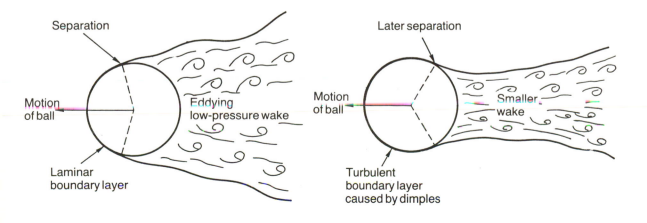

itself, and the additional energy associated with such random motion
enables the boundary layer to progress further towards the back
before separating from the surface. The result is a significantly smaller
low-pressure wake and a corresponding reduction in drag. Best results
are achieved with dimples about 0.5 mm deep.

Golfers have also discovered how to create lift as the ball flies
through the air. They do it by striking the ball so as to impart back-
spin – upper surface spinning backwards, lower surface spinning
forwards – up to many hundreds of revolutions per minute. To
understand how this works, imagine a golf-ball suspended in a wind-
tunnel and back-spinning about a horizontal axis as shown in Fig.
5.20(a). As the air flows over the top surface, its speed is increased
because of the spin, whereas the opposite happens at the lower
surface. Higher speeds correspond to lower pressures and the net
effect is an upwards lift force which can lengthen the drive
considerably.

The aerodynamics of cricket balls is a subject in which it is so
difficult to separate fact, verisimilitude and myth that it will continue
for the foreseeable future to engage the attention of serious-minded
researchers as well as of expert players. The rôles of speed, seam,
surface, spin, humidity, temperature, pressure and even cloud cover
all contribute to cricketing lore, but although some of the finer points

Fig. 5.20 *The advantage of spin
(a) in golf (b) in cricket.*

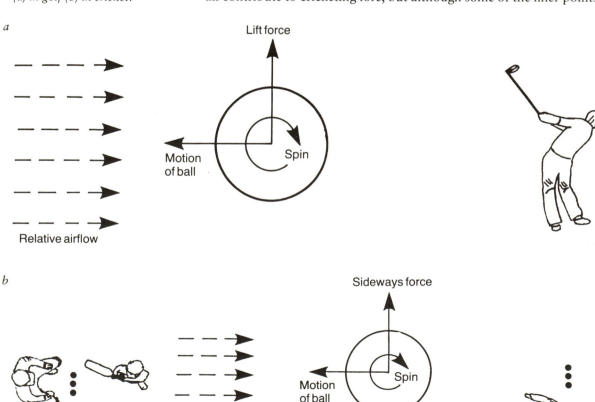

are still arguable certain effects can confidently be attributed to their causes.

Consider, for example, the effect of spin on the movement of the ball through the air, Fig. 5.20(*b*). Spin bowlers deliver the ball at fairly modest speeds of 30–40 miles per hour, but compensate for the length of the advance notice given to the batsman by making the ball move sideways in flight in the hope of confusing the waiting striker. Their technique, like that of a golfer, depends on Bernoulli's famous law. Consider what happens when the bowler spins the ball about a vertical axis as it leaves his hand. If the direction of spin is clockwise looking down, the speed of the air relative to the centre of the ball is increased on the bowler's right and decreased on his left, with the result that a lower pressure is developed on the right than on the left – and the ball will swerve to the right as it proceeds down the wicket. The opposite will happen if the spin is reversed. Moreover, if the axis of spin is not vertical but inclined, various combinations of flighting together with sideways movements are possible.

Fig. 5.21 shows a large model of a cricket ball suspended as a pendulum on a light rod. It was attached to the rod such that it could be spun freely about the axis of the rod and, when placed in a stream of air giving about the same Reynolds number as a spin bowler gives a real cricket ball, it deflected sharply to one side when spun.

Fast bowlers who deliver at over 70 miles per hour work differently. At those speeds, the boundary layer remains laminar around the front of the ball, but separates at maximum diameter to leave behind a turbulent low-pressure wake, which in turn creates a large drag on the ball, rather like the smooth golf-ball of Fig. 5.19(*a*). At somewhat higher speeds, the picture changes to look like the dimpled golf-ball of Fig. 5.19(*b*). The front side boundary layer becomes turbulent itself, which, as we have seen, enables it to stick to the surface more successfully; separation occurs further back and both wake and drag are reduced. This by itself is a remarkable effect. If the

Fig. 5.21 *A large version of a cricket ball in a stream of air flowing from left to right. When spun clockwise looking down, the ball deflected sharply in the direction shown.*

bowler can get *above* the critical speed for the first few yards of his delivery, then as the ball decelerates there will be a sharp *increase* in drag and the flight path will dip in a seemingly unnatural way, clearly a useful result from the bowler's point of view. The numbers involved lend support to these ideas: at speeds just above 70 miles per hour, the drag is about equal to one-quarter of the weight of the ball, but at speeds just below this critical value it becomes about equal to the weight of the ball itself.

But that is not all. Although the ball is nearly spherical, its surface is interrupted by the seam, a zone about 0.75 in (19mm) wide with stitches standing up by 0.02 in (0.5mm) on a new ball, about the depth of the boundary layer. This gives an expert bowler an opportunity to swing the ball in flight. If he delivers the ball with the plane of the seam inclined to the direction of forward motion, as shown in plan in Fig. 5.22, and, if the forward velocity is just below the critical value, the seam at the front will trigger turbulence in the boundary layer and delay separation on that side, whilst on the other side separation will occur as usual at about maximum diameter. The wake and the corresponding pressure distribution are no longer symmetrical, and the resulting force on the ball will make it swing sideways. The force involved can be considerable, not much less than the weight of the ball, so that the sideways swing can be of considerable help in finding the edge of the bat. Assiduous polishing of one side of the ball also helps to increase the asymmetry of the flow and hence the swing but, if the ball has become excessively rough, the seam loses its advantage because the boundary layer will be turbulent at high speeds anyway. And in order to keep the seam reasonably stable in the chosen direction, the bowler has to impart a small amount of spin in the plane of the seam. It is less surprising that only a few bowlers can get it all right than that any can do it at all.

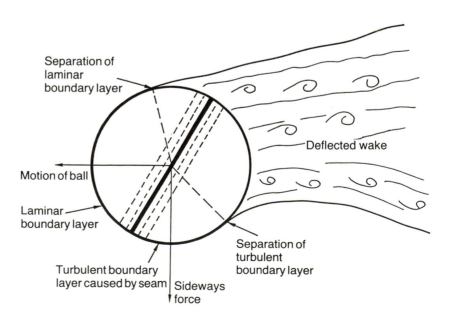

Fig. 5.22 *The aerodynamics of swingers, plan view.*

Fig. 5.23 *The Flettner rotor ship
of 1925.*

The Flettner rotor-ship, which aroused intense interest by crossing the North Sea from Gdansk to Grangemouth in 1925 and the Atlantic a year later, was an extraordinary vessel which employed spin to generate a propulsive force from the wind. As shown in Fig. 5.23, it had no sails, but it was nevertheless a sailing ship, albeit equipped with an auxiliary engine. The two tall cylinders, which were driven at modest speeds about their vertical axes by below-deck machinery, were the key features of the design. To fix ideas, imagine that the ship is pointing due west and the wind is blowing from the north. If the cylinders are rotating counter-clockwise looking down, then the air passing over the front of the cylinder will acquire a higher velocity (and therefore a lower pressure) than the air on the other side, and a net propulsive force is available to drive the ship. Clearly, the forward motion of the ship affects the matter; but, if the wind velocity relative to the ship is considered instead of the absolute wind velocity, the same conclusion applies. Moreover, if the relative wind were blowing from the south, a forward force could still be generated if the cylinders were driven in the opposite direction.

Rotating fluid machines

A moment's reflection on the subject of rotating fluid machines is enough to convince anyone of their value. From the water-wheels and windmills of ancient times to the enormous variety of fans, blowers, pumps, hydrodynamic transmissions, propellers, helicopter rotors, compressors and turbines (water, steam and gas), the range is not only a tribute to mechanical ingenuity but also an essential underpinning of modern civilization. The design and manufacture of each type of machine demands its own hard-won know-how, and for the engineer the advances made possible by current research present exciting prospects for the future.

As we saw in previous chapters, a large proportion of the world's electricity supply is provided by steam turbines driving electrical generators. Fig. 1.1(*c*) gives an impression of the scale of a 660 MW turbine. As the steam travels through the turbine, it passes progressively through groups of wheels in the high-pressure, medium-pressure and low-pressure cylinders, at the end of which its pressure has dropped from nearly 160 bar to less than 0.05 bar and its volume has increased a thousand-fold. Fig. 5.24(a) gives an indication of the size and the engineering quality of low and high-pressure blades. Fig. 5.24(b) shows a view of how the steam is directed to the blades by a matched ring of nozzles fixed to the casing and interposed between one wheel and the next, each combination of one fixed and one moving ring being known as a *stage*.

The detailed design of the nozzles and the blades is a matter of great importance for the performance of the turbine; it has to account for variations in pressure, volume, temperature, wetness and velocity of the steam as it passes through successive stages. As there are many stages arranged in series, each depending on the exit conditions of its

a

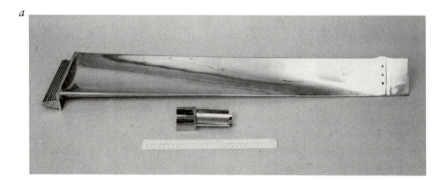

b

Fig. 5.24 (a) Low- and high-pressure blades of a large steam turbine, and (b) positioning a low-pressure section of the rotor in the casing, where the nozzles fixed to the casing can be seen at the bottom of the photograph. Courtesy of NEI Parsons.

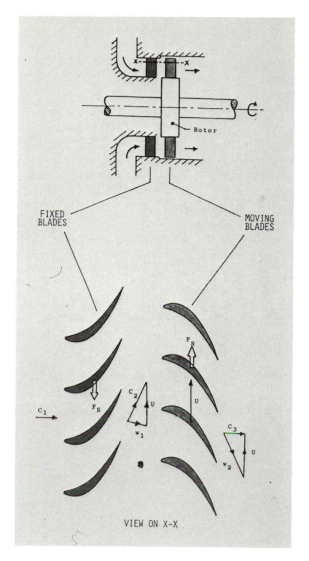

Fig. 5.25 *Flow through a single stage of a steam turbine.*

predecessor, it is essential to ensure that each contributes its expected part to the whole.

Fig. 5.25 shows a simplified version of the flow through a single stage. When the predominantly axial flow of steam encounters the fixed blades forming the nozzles, it is both accelerated and deflected by the shape of the blades. This increases its speed as it meets the moving blades and also directs the flow so that it enters the passages between the moving blades at the correct angle. The velocity v of the moving blades enters the account at this point, because the decisive factor is not the absolute velocity of the steam at entry, c_2, but rather its velocity relative to the moving blades, v_2. With this in hand, the steam passes over the surfaces of the moving blades, from which, after a second round of acceleration and deflection, it emerges with a lower pressure, a larger volume and once again a predominantly axial velocity c_3. The changes that take place in that second round determine the force which the steam exerts on a moving blade, and the shape of the blade is determined largely by this requirement. Inevitably, as the condition of the steam varies greatly from one end of the turbine to the other, the shape of the blades, both moving and fixed, also varies greatly, but as might be expected a distinct resemblance to aerofoil shapes is apparent in all cases.

Unlike *steam* turbines, which are driven by steam generated by heating water in a boiler, *gas* turbines eliminate the intermediate stage of steam raising by using the gaseous products of combustion directly to drive the rotors.

One of the earliest designs, due to John Barber of Nuneaton (1791), Fig. 5.26, involved a heated retort generating combustible gas which was mixed with air and ignited before issuing at high speed through a nozzle to strike a primitive turbine wheel. Since the 1940s, the gas turbine has developed mightily to provide the power for a large

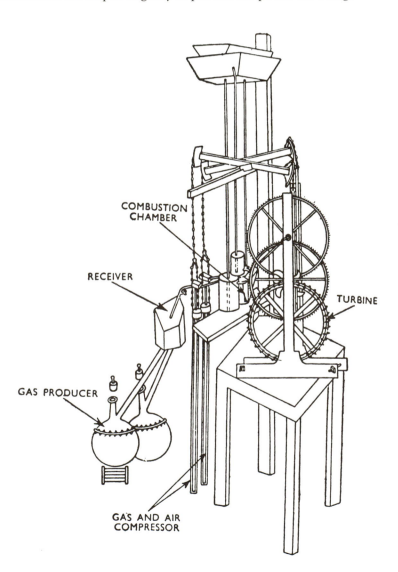

Fig. 5.26 *The first known proposal for a gas turbine, by John Barber in 1791.*

proportion of the world's aircraft, many ships and land vehicles, and a host of industrial applications. Its operation is similar to that of a steam turbine in the sense that the working fluid driving the rotor must expand from high pressure at inlet to low pressure at exhaust; but, because it has no independent source of high pressure, the gas turbine must incorporate a means of generating high-pressure gas on its own account. This is provided by a compressor, driven by the turbine itself, which sucks in the incoming air and compresses it before delivery to the turbine. Naturally, even if there were no losses in such a combined system, the surplus power available at the turbine-shaft would be precisely zero; all would be absorbed in driving the compressor. Clearly, to get power out, something else must be done. The essential addition is to burn a supply of fuel in combustion chambers interposed between the compressor and the turbine. Both temperature and pressure of the resulting air–fuel mixture can then be raised to levels which enable the turbine not only to drive the compressor but also to deliver shaft-torque or jet-thrust as required.

Fig. 5.27 illustrates a simple lay-out in which the shaft of a single turbine is mechanically coupled to the compressor shaft. Many other more elaborate arrangements have been developed for particular applications, including the recovery of heat from the high-temperature exhaust gas leaving the turbine to heat the air flowing between the compressor and the combustion chambers, the division of the turbine into stages with additional combustion chambers fed by an independent fuel supply between stages, and so on. And many factors, such as weight, type of fuel, required ranges of load and speed, running costs, initial costs and life have to be taken into account in arriving at the best design for the purpose.

In aircraft applications, the gas turbine is used to supply shaft-torque to a propeller for flying speeds of a few hundred miles per hour, or as a pure jet engine for speeds above about 500 miles per hour when the propulsive efficiency becomes superior to alternative forms. Recent advances in aviation owe much to the spectacular progress that has been made in the design of gas turbines for jet

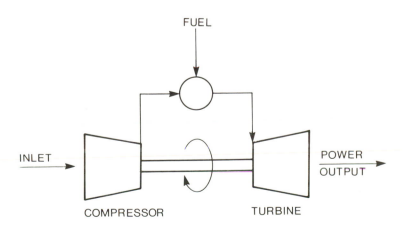

Fig. 5.27 *Layout of a simple gas turbine.*

engines. Their purpose is to generate a thrust by receiving air at the inlet, and, after due process in the engine, emitting a high-velocity jet at outlet. Fig. 5.28 shows a version of the Rolls-Royce Ardour engine as used in many operational aircraft; it can deliver nearly 6000 lbf (26 688 N) of thrust, a figure that can be increased by afterburning of additional fuel in the exhaust stream. Two gas turbines are incorporated in the design – a two-stage low-pressure compressor driven by a single-stage low-pressure turbine, and a five-stage high-pressure compressor driven by another single-stage high-pressure turbine.

Details of the drawing show that the comon shaft of the first combination is nested inside the shell-like shaft of the second. A proportion of the airflow leaving the low-pressure compressor is ducted past the combustion chambers in a bypass arrangement, thereafter joining the exhaust from the low-pressure turbine. Such a bypass confers several advantages. If a propulsive efficiency is defined as the ratio of power delivered to kinetic energy available in the gas-flow, then for a *non*-bypass engine, the efficiency is proportional to $(v_j - v_0)v_0/(v_j^2 - v_0^2) = 1/[1 + (v_j/v_0)]$, where v_j is the velocity of the jet and v_0 the airspeed. It pays therefore to keep v_j low compared with v_0 – but for high thrust this would mean larger engines. By mixing two streams of flow as shown, the bypass avoids this dilemma by reducing the joint jet velocity and still maintaining thrust. It also leads to a quieter engine.

Fig. 5.28 *Section through a Rolls-Royce Ardour jet engine.*

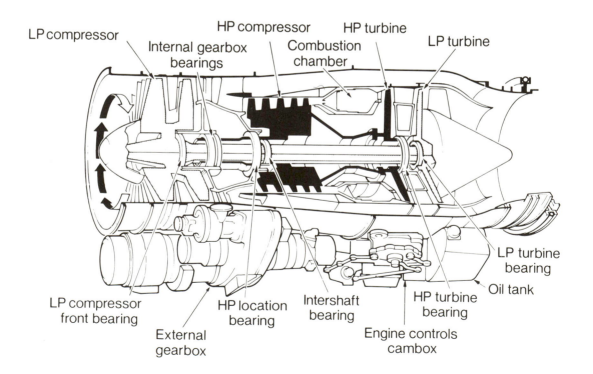

Faster than sound

So far, we have been thinking of fluid flows as if the fluid were incompressible. Although this applies widely to liquids, it is of great practical interest for gas-flows to know when compressibility becomes important and what are its effects.

The broad answer to the first question is simple: compressibility becomes important when the speed of the fluid approaches the speed of sound in the fluid, or when the ratio (v/c) approaches unity, where v represents the speed of the fluid and c that of sound in the fluid. This ratio is called the Mach number M in memory of the great aerodynamicist Ernst Mach. For an ideal gas, the speed of sound can be expressed very simply as

$$c = \sqrt{(\gamma R T)},$$

where γ is the ratio of specific heats at constant pressure and constant volume, R is the universal gas constant, and T is the absolute temperature. For air of moderate humidity, which approximates closely to an ideal gas, the result at 15°C (absolute temperature 288 K) is 340 m/s or 733 miles per hour.

The answer to the second question is far from simple. Thermodynamic properties of gases – the dependence of density on pressure, of internal energy on temperature and so on – play an essential rôle in the proceedings, and the ways in which changes occur – whether with heat transfer or friction, for instance – must also be taken into account. Some examples may serve to illustrate particular effects.

Consider a thin aerofoil at a small angle of attack in an airstream flowing past at $M = 0.5$, half the speed of sound. Theory shows that the lift calculated on the basis of compressible flow is about 16% higher than that calculated by assuming incompressible flow. If $M = 0.7$, which is close to the limit when the flow at some point around the airofoil becomes sonic, the difference increases to about 32%. At higher speeds, a significant part of the flow becomes supersonic, and the entire flow pattern changes.

A striking effect of compressibility occurs when a gas flows in a convergent straight duct, Fig. 5.29. If at first the gas is assumed to be incompressible and all variations in density are ignored, then because the rate of mass flow through each section must be the same, the product $(\rho A v)$ of density (ρ), area of duct (A) and gas velocity (v) must be equal at all sections. If ρ remains constant, a decrease in A must inevitably be accompanied by an increase in v. If, however, we admit that density changes with pressure, and recall that pressure depends on velocity, the position is not so simple. The same mass flow must again pass through all sections, but now even larger velocities occur because of the drop in density as the gas accelerates. This process will continue until a critical velocity – the velocity of sound – is reached. If the duct continues to converge after that stage, which in incompressible flow would lead to still higher velocities, further reductions in pressure can no longer be propagated upstream into the

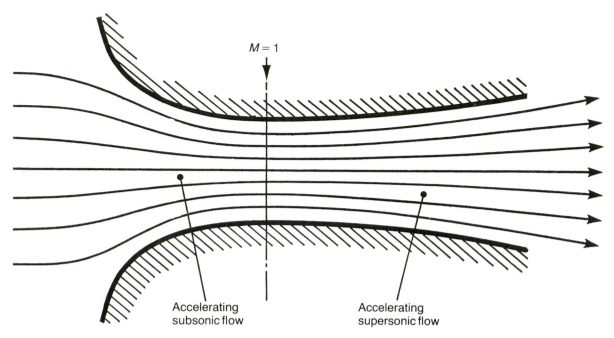

$M = 1$

Accelerating
subsonic flow

Accelerating
supersonic flow

Fig. 5.29 *Accelerating flow in a convergent–divergent duct having sonic velocities ($\mathbf{M} = 1$) at the throat.*

approaching gas so as to maintain steady conditions, and the flow becomes choked.

Surprisingly, however, further acceleration is possible if from that critical section onwards the duct is made to *diverge*. At supersonic speeds, an increase in flow area A causes such a rapid fall in density ρ that the product (ρA) falls despite the increase in A. The only way that the mass flow rate $(\rho A v)$ can remain constant – as it must – is for the velocity v to increase, and the remarkable result follows that velocity in a diverging duct will continue to increase supersonically. This reversal of ordinary expectation, illustrated in Fig. 5.29 forms the basis of convergent–divergent ducts of rocket engines from which the jet issues at supersonic speeds, and in windtunnels where a stream of air has to be accelerated from rest to supersonic velocities in a test section.

The sounds we hear are created by small variations in air pressure which impinge on our ears and send electrical signals to the brain via tiny movements of the ear's complicated structure, and the speed at which they travel through the air plays a central rôle in compressible gas dynamics.

Fig. 5.30 shows a small source of sound which emits bleeps of sound at regular intervals into the surrounding stationary air. In (*a*), the source is stationary, too, and the progress of successive pressure variations is marked by increasing spherical surfaces called sound waves, which appear as circles in our two-dimensional picture. In (*b*), the source is moving to the left at half the speed of sound. At the current position D, the sound wave has had no time to leave D; however, the sound wave emitted one second earlier at position C has reached the smallest circle, the sound wave emitted two seconds earlier at B has reached the next circle, and so on. The sound waves progressively out-distance the source, and the stationary air in front of

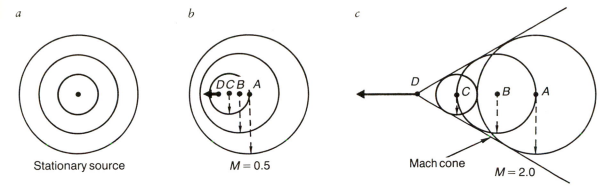

Fig. 5.30 *A pattern of sound waves. In* (a), *the source of sound is stationary; in* (b), *it is moving to the left at half the speed of sound* ($M = 0.5$), *and the circles show how far the sound wave has travelled from successive positions of the source; in* (c), *the source is moving at twice the speed of sound* ($M = 2.0$) *and outside the Mach cone there is a zone of silence.*

the source continues to receive advance notice of its progress. The position is entirely different if, as in (c), the source is travelling faster than sound, say twice as fast. Then the sound emitted at C has been able to travel only half the distance moved by the source, the sound from B has fallen even further behind, and no sound whatever has reached the air outside a cone marked by the two straight lines. Outside the cone there is silence.

When a solid body moves through the air, the variations of air pressure caused by the motion are much larger than those caused by sound, but their behaviour follows the same general pattern. For aerodynamic purposes the body is taken to be stationary in a flowing stream of air, rather than the other way round, and Fig. 5.31(a) illustrates such airflows past an inclined flat plate when the flow is (a) subsonic and (b) supersonic. To simplify matters, we imagine that the plate is infinitely long in a direction perpendicular to the diagram, so that the flow is two-dimensional, and ignore the complications of friction and heat transfer from air to plate.

In the subsonic flow of (a), the streamlines pass through regions in front of the plate where the pressure variations make them deflect in a smooth continuous fashion to accommodate the obstacle ahead. In the supersonic flow of (b), the streamlines have no such advance notice. Nevertheless, they must follow the inclined surfaces of the plate, and changes in velocity and pressure must occur somewhere along their paths. Like the discontinuities of the moving source of sound, these changes occur suddenly when the streamlines encounter the shock waves shown by the inclined lines at the leading edge of the plate. Similar shock waves attached to the trailing edge of the plate restore the free stream conditions of flow. The details present a considerable mathematical challenge. When real physical features such as finite wing-spans, aerofoil shapes, boundary layers and viscous effects are taken into account, the design task is formidable. But it is easy to see that substantial changes in lift and drag will occur during transition from subsonic to supersonic flight, and it is a measure of the success of aerodynamic design that such a transition can be accommodated without loss of performance. Fig. 5.31(b) shows actual shock waves observed in a wind-tunnel.

Anyone who has drawn a stick through a pool of water will have noticed the surface waves that fan out behind it. Although they differ

Fig. 5.31 (a) *Subsonic and supersonic flows past a flat plate. In the subsonic flow, the streamlines bend smoothly around the plate, but in the supersonic flow they change suddenly as they encounter the discontinuities of the shock waves. (b) Shock waves observed in a supersonic wind-tunnel at M.I.T. Gas Turbine Laboratory. Courtesy of Ronald Press.*

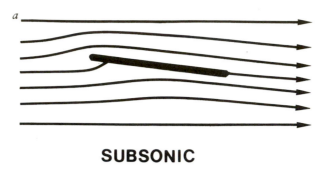

SUBSONIC

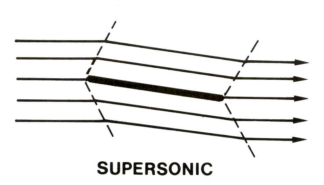

SUPERSONIC

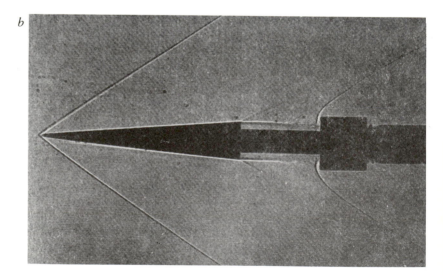

physically from aerodynamic shock waves, their underlying mathematics is so similar that useful comparisons are possible. Fig. 5.32 shows an experiment in which a thin sheet of water flowed down an inclined plate past a thin diamond-shaped obstacle placed in

its path. Soon after the flow was released, several stationary surface
waves became visible. The most pronounced were generated by the
sharp leading edge of the diamond, but others were attached to the
mid-length corners and it was even possible to see waves reflected
from the sides of the container. This pattern corresponds closely to the
formation of shock waves in supersonic gas-flows, and the analogy
can provide useful insights into phenomena of much greater
experimental difficulty.

Although elaborate apparatus is required to *see* a shock wave, it can
be *heard* in a simple and familiar way. If a long whip is flicked
through the air, a small piece of material at the tip can be made to
travel at a supersonic speed and to generate a modest and short-lived
shock wave. When this was attempted with a whip made of stranded
cord knotted at the tip to improve matters, sharp cracks were heard as
the shocks were propagated through the surrounding air, and small
pieces of cord, stripped from the tip of the whip by the sharp changes
of pressure, floated gently to the ground.

Fig. 5.32 *Surface waves created
by a flow of water past a sharp-
edged obstacle.*

Full circle

Anyone interested in exploring the curious behaviour of a boomerang
can do so by making one out of two wooden rulers, as shown in
Fig. 5.33(*a*). Better results can be obtained if the cross-sections of the
wood are shaped in the manner of aerofoils, thicker near the outside
than the inside edges as shown in the miniature but real boomerang,
but with luck even rulers will do. According to the ideas of Chapter 2,
gyrodynamic stability of spin in the plane of the boomerang is assured
if the angle between the two arms is close to a right angle, but the
experimentalist should be warned that small changes in construction

Fig. 5.33 (a) *Boomerangs,*
permanent and temporary; (b) an
expert throw.

a

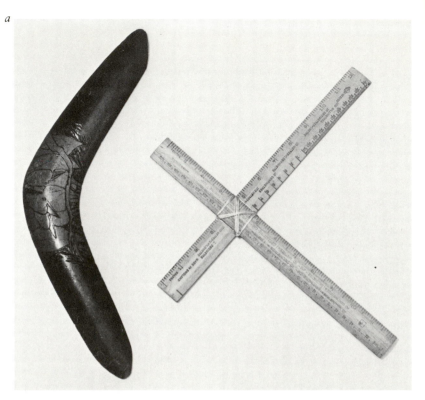

b

– length of arms, position of cross-over, tapering sections and so on – are likely to offer marked variations in performance.

The method of launch is to grip the end of one arm and to throw the boomerang forwards with a rapid forward spin (top side faster than bottom side) in a plane slightly inclined clockwise from the vertical as viewed by the thrower. The aerodynamic 'lift' forces on the blades act at right angles to the plane of the launch but more strongly on the advancing blade than on its retreating companion (Bernoulli again). If, as supposed, the cross-sections of the blades are convex on the thrower's left, the resultant sideways force averaged over a revolution will also be to the left, and the path of the centre of gravity G will curve in that direction.

But as well as affecting the *path* of G, the resultant lift force also supplies a *torque* about G, to which the spinning boomerang responds by precessing in the manner of the gyroscopes described in Chapter 2. One component of the torque acts about an axis pointing back towards the thrower, and so causes the plane of spin to turn slowly about a vertical axis to align with the forward motion of G, rather like a steered car wheel. The second component arises because the average (left-pointing) lift force lies *ahead* of G; this creates a torque about a *vertical* axis and makes the plane of spin tilt or precess towards the horizontal – the boomerang 'lies down'. As the tilt progresses, the lift forces point increasingly upwards and help to keep the boomerang airborne as it continues on its curved flight.

There are, of course, many more mysteries in the matter, but they should serve only to encourage the enthusiastic investigator. Fig. 5.33(*b*) shows the progress of a near-perfect flight achieved with crossed sticks in the lecture theatre.

6

LIVING MACHINES

An infinite variety of motion

Zoologists estimate that nature has been developing life-forms on Earth for about 1000 million years. Animal movements are astonishingly varied but all are subject to the same physical laws of motion; nothing has yet been discovered in the life-sciences to upset the general principles of dynamics. But, in view of the differences

Fig. 6.1 *Two of the many drawings of natural motions by Leonardo da Vinci.*

a

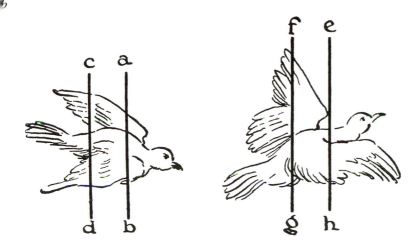

between life-forms on the one hand and machines on the other, it would be rash to expect that the laws can be transferred without complication from one field to the other.

Nevertheless, there is much fertile common ground and it offers not only a better understanding of how things move but also valuable practical benefits. For example, investigations of the dynamics and control of animal movements, which involve a rich variety of feedback mechanisms, have led to new prospects for machines such as robots, whilst in the field of aerodynamics devices of great interest have been discovered through the study of airflows over birds' wings. Starting from the opposite quarter, engineering design applied to medicine has helped to achieve successes that were inconceivable a few years ago; the repair and replacement of human joints and organs, and the development of new prosthetic aids for the disabled, illustrate how the transfer of ideas across traditional boundaries has contributed to outstanding practical successes.

As usual, the genius of Leonardo da Vinci opened up the field of study dramatically. Two of his many drawings of natural motions are reproduced in Fig. 6.1. A variety of other drawings of different eras, all directed to a better understanding of the complicated motions of living machines, are shown in Fig. 6.2.

A striking difference between man's and nature's machines is that there is nothing in the latter that constitutes a wheel. The wheel-and-axle is an entirely human design, some 6000 years old. It has been developed progressively in machines of many different types, and without it few of them would be able to operate at all. Its success depends on sliding or rolling contacts between adjacent parts, which allows one part to keep on turning indefinitely relative to another. In the animal world, the connections between living parts have to ensure an uninterrupted supply of blood and a continuous connection of nerves. This poses a problem of mechanical design so difficult that nature has apparently ruled out indefinite rotation as a possibility: instead, *limited* rotations, in which the vital connections are

Fig. 6.2 *Views of natural motion:*
(a) and (b) from Borelli,
17th century; (c) galloping horse
and cheetah (from Hildebrand,
1959); (d) leaping frog (from
Gray, 1968).

a

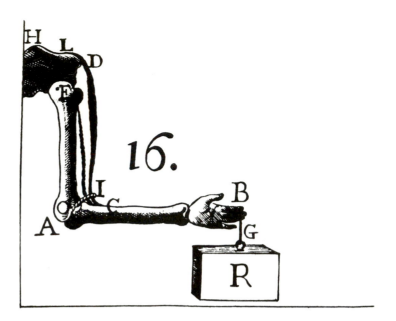

b

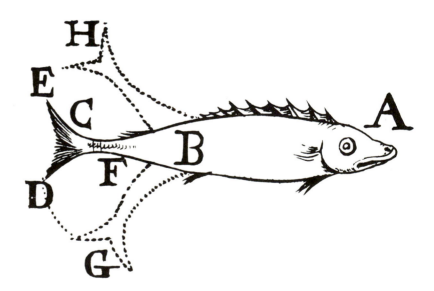

maintained without separation, have been developed to provide the
amazing repertoire of animal movements. An alternative explanation
accepts that the process of evolution is perfectly capable of developing
sliding contacts, such as retractable claws (which do not require the
supply of blood and nerves), but that reciprocating legs offer a better
means of locomotion than wheels on land surfaces without hard
roads. In water or air, oscillating bodies or wings use the available
energy at the speeds in question more efficiently than screws or
propellers. In other words, under the prevailing conditions, nature
found *better* solutions than the wheel.

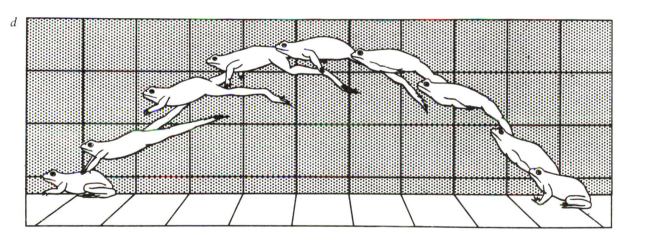

A simple demonstration of limited joint rotations can be performed by extending an arm straight forward from the shoulder in a horizontal position with the thumb pointing upwards as a marker. It is an easy matter to move the arm continuously in a wide curve about that central position while keeping the thumb pointing upwards. Clearly the thumb travels continuously in a circle, which suggests a continuous rotation, but a moment's thought will confirm that the thumb is not rotating at all – on the contrary, it always points in the same direction. What *does* rotate is an imaginary line drawn from the thumb to the centre of the circle, but that, of course, does not imply a continuous rotation of the shoulder joint. Although the capacities of individuals joints, such as the ball-in-socket joint at the hip and the hinge joint at the elbow, have been well-documented, cooperative rotations of a set of joints arranged in series, as from the shoulder to the wrist, pose difficult problems of three-dimensional dynamics and control.

Crawlers

Without the wheel, the animal world has developed many ingenious ways of moving on dry land. The leech, for example, fastens down a suction pad at the rear, stretches itself so as to move its head forwards, fastens down another suction pad at the head, then releases the rear pad and contracts to complete a cycle of forward motion. The earthworm adopts a similar pattern of movement but, being longer, can perform more versatile manoeuvres involving sideways bending and pushing against any convenient fixture that comes to hand.

Fig. 6.3 shows the projection of an earthworm moving over the flat top of an overhead projector. In the position shown, the worm had advanced from the smooth surface on the right-hand side into a region where small discs stuck on the surface provided useful purchase. After a tentative exploration of the left-hand side, where the surface had been deliberately roughened, the worm found it less to its liking and beat a hasty retreat.

Snakes exhibit similar but more effective means of locomotion. Anyone who has watched a snake gliding rapidly forward must have been struck by the resemblance to the flow of water along a narrow undulating channel; the motion is continuous, each element of the snake following the path of its predecessor in a rapid, graceful glide. Such serpentine motion, illustrated in Fig. 6.4(*a*), depends on the transverse forces which the snake can develop against convenient fixtures, or against the indentations made by its own weight on a soft surface. Internally, the muscles of the snake pull or push the small axial elements of its skeleton along the desired path. An alternative method of propulsion, aptly called concertina motion, finds favour when the variable undulations in the path are not so convenient or, as in a straight channel, absent altogether. Rather more dynamics is then required of the snake as it seeks alternative fixtures against which to push. The resulting motion is illustrated in Fig. 6.4(*b*), where by

Fig. 6.3 *An earthworm in transit over an overhead projector.*

a

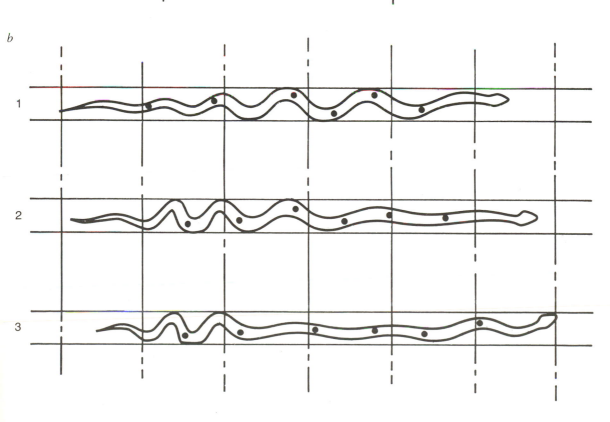

Fig. 6.4 *Progress of a snake:*
(a) *serpentine,* (b) *concertina,*
from Gray, 1968.

b

Fig. 6.4 (c) *Side-winding,*
(d) tracks left in the sand as the
side-winder moved from left to
right, with the head at the bottom
of the photograph. (c) from Gray,
1968; (d) from Klauber, 1944.

forming bends which lock against the sides of the channel, the
remainder can either be pulled or pushed forwards, depending on the
internal position of the fixed bend.

A third more complicated pattern known as side-winding is
practised by rattlesnakes and is illustrated in Fig. 6.4(c). The general
motion is achieved by first setting down two fixed elements which
push against the ground, and then progressively lifting the remainder
upwards and sideways until two new anchors are secured somewhat
nearer the tail. By the time the rear-most anchor has reached the tail,
the head has advanced sufficiently to provide a new anchor, and the
process is repeated. Segments of track marked by the forces exerted on
the ground at the points of contact are shown in Fig. 6.4(d). Fig. 6.5
shows an Indian rock-python from which the lecturer was trying to
disentangle himself without any particular reference to the motions
described above.

A walker

The walking and running of animals have been studied extensively,
and in a later section we shall be looking at aspects of human

Fig. 6.5 *Unclassified motion of an Indian rock-python.*

locomotion. But before we leave dry land, a glance at the extraordinary feats (!) of the myriapod, illustrated in Fig. 6.6, reveals an almost unimaginable degree of coordination and control. In ordinary walking, a regular wave pattern passes along the large number of limbs from end to end, successive groups providing both

Fig. 6.6 *An annoying loss of contact and coordination, from Gray 1968.*

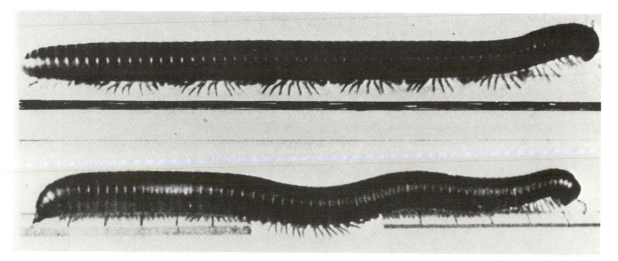

support and propulsion in a highly synchronized way. When passing over a gap, coordination is lost, but regained on returning to *terra firma*.

Swimmers

The swimmers show even more variety than the crawlers. Floating rowers, such as swans, ducks and geese use their legs as oars, while their relatively rigid bodies conceal the brisk pedalling going on below. Their webbed feet are well-suited to develop thrust on the back-stroke, and to minimize drag as they draw their legs forwards again for the next round of propulsion. Water-beetles, too, use their legs for rowing (Fig. 6.7) and, with three pairs of legs available, have developed the rear pair as the main propulsive machinery; not only are they articulated for extension on the backwards power stroke and flexure on the forward, but they are also equipped with fine transverse hairs which improve the hydrodynamics of flow.

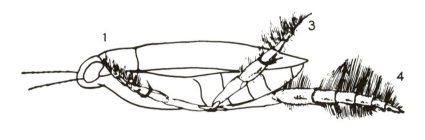

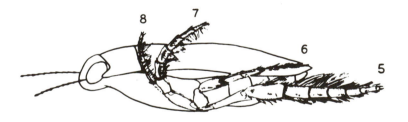

Fig. 6.7 *Rowing of a water-beetle. The numbers indicate successive positions of the left hind leg: the other two left legs and all three right legs have been omitted. From Nachtigall, 1960.*

Squids, jelly-fish and octopuses rely on jet propulsion for fast spurts. The squid takes on board a quantity of water through the space between the mantle and the body, Fig. 6.8(*a*), then by closing the mantle around an interior funnel squirts it out again to generate momentum in the opposite direction. In the position shown, the funnel is pointing forwards to produce a rapid acceleration backwards, but it can be turned into various other directions according to the direction in which the squid wishes to go. The jelly-fish operates on a similar principle but by less elaborate means, Fig. 6.8(*b*).

A third type of propulsion resembles flying in that fins moving through the water generate both lift and thrust. Fig. 6.9(*a*) illustrates the action of an emperor penguin swiming under water and beating its wings like a bird; the wings act as hydrofoils and are adjusted during their up-and-down movement in such a way as to provide thrust by

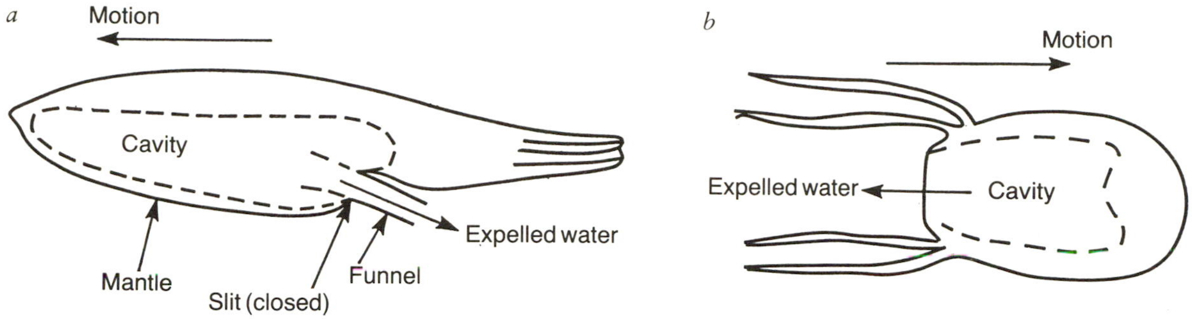

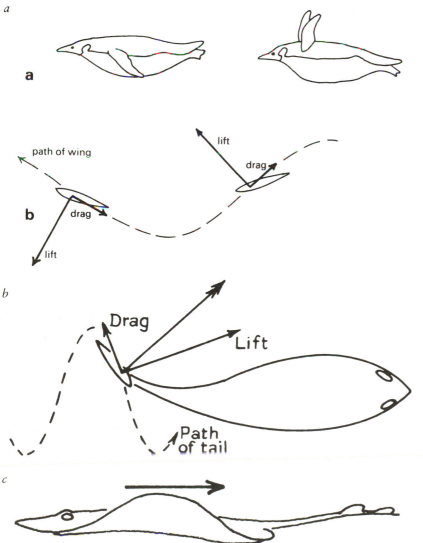

a

Motion

Cavity

Mantle

Slit (closed)

Funnel

Expelled water

b

Motion

Expelled water — Cavity

means of a forward component of lift. Whales swim by oscillating
their horizontal tail flukes up and down to produce a similar hydrofoil
action, and tunny fish, Fig. 6.9(*b*), oscillate their stiff vertical fin at the
tail. The fins of sharks, vertical at the tail and horizontal along the
body, also act as hydrofoils, although their swimming action also

Fig. 6.8 *Jet propulsion as
practised* (a) *by a squid,* (b) *by a
jelly-fish.*

a

a

path of wing

lift

drag

b   drag

lift

b

Drag

Lift

Path
of tail

c

Fig. 6.9 *Swimming by hydrofoil:*
(a) *an Emperor penguin under
water (from McNeill Alexander,
1982);* (b) *plan view of a tunny
(from McNeill Alexander and
Goldspink, 1977);* (c) *a ray,
showing the direction of the
undulations passing along the fins
(from McNeill Alexander, 1982).*

depends on undulations of their bodies as a whole. One of the most tranquil swimmers is the ray, Fig. 6.9(*c*), which is equipped with large horizontal fins roughly in the shape of triangles; here too the fins act as hydrofoils, but in addition undulations passing backwards along the fins reinforce the forward thrust.

Many studies of undulatory modes of swimming have been carried out in recent years as fluid dynamicists have joined forces with zoologists in exploring the complicated phenomena involved. Where the undulations involve flexing some or all of the body, the hydromechanics are of particular importance because of the many orders of fish that depend on it for propulsion. They are usually grouped according to the extent of the flexure along the length of the body. In the first category are those slender eel-like swimmers which flex from head to tail; in the second, including the trout and the barracuda, most of the flexure is confined to the rear part of the body; and in a third group, the swimmers keep their bodies relatively rigid but oscillate stiff fins and tail like activated rudders. Fig. 6.10 shows examples of the three forms.

Although the hydromechanics of propulsion differs significantly in the three groups, the underlying principle of pushing on the surrounding water by elements of the body is the same. In Fig. 6.11(*a*) a trout is shown propelling itself in air over the surface of a board fitted with smooth pegs; being substantially normal to the surface of

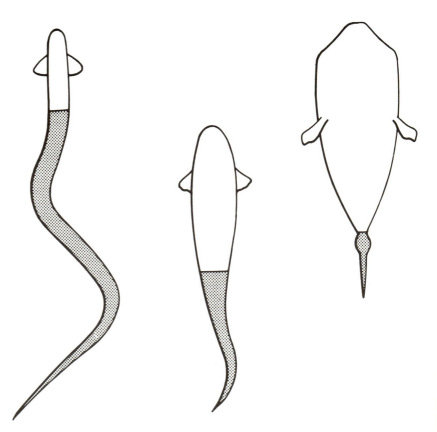

Fig. 6.10 *Three types of aquatic propulsion. From Gray, 1968.*

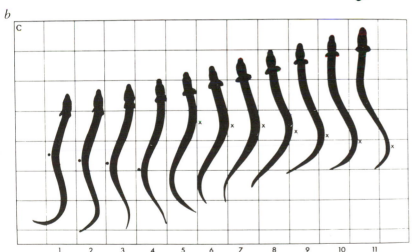

a

S_1

S_2

S_3

S_3

b

C

1   2   3   4   5   6   7   8   9   10   11

Fig. 6.11 (a) *A trout propelling itself in air over a board fitted with smooth pegs;* (b) *a butterfish swimming in water. From Gray, 1968.*

the fish, the forces exerted by the pegs provide the forward components required for propulsion. In swimming, of course, the elements of water in contact with the fish move as the body presses against them, and such motion must be taken into account in analysing the resulting hydrodynamic forces. Fig. 6.11(*b*) shows successive positions of a swimming butterfish marked at 0.5 s intervals; the grid is one inch square and the positions are drawn displaced progressively to the right for clarity of presentation. It can be seen that the whole body participates in organized transverse movements – larger near the tail, smaller near the head – and that the crests of the waves move backwards with respect to both the fish and the gridlines. Although the fluid mechanics of such flexible motion in water is complicated, mathematical methods of analysis have made rapid progress in discovering the precise mechanisms involved. And the shapes of certain fish confirm the salient features, such as low drag, of efficient aerofoils, as illustrated in Fig. 6.12.

Fig. 6.12 von Kármán's comparison of Cayley's measurement of the body of a trout with NACA low-drag aerofoils: circles refer to the trout, full and dashed lines to NACA 63 A016 and LBN 0016 sections, respectively.

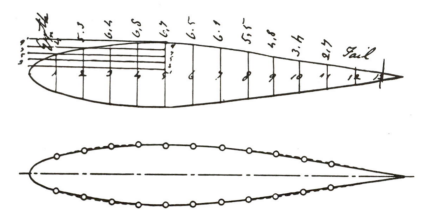

Flyers

Unlike most fish in water, birds are not buoyant in air and so the generation and control of lift in flying is a matter of primary importance.

Gliding flight is illustrated in Fig. 6.13. The wings of the bird passing through the air create lift and drag forces dependent on speed in a manner similar to the action of aerofoils. In still air, the drag force, composed of frictional resistance and induced drag associated with lift, would soon cause a rapid deceleration and hence loss of lift unless the forward velocity were sustained by losing height. The resultant flight path is therefore determined by the balance achieved amongst the various factors involved, including the angle of attack of the wing, its surface area and the wing configuration assumed by the bird. By means of wing control, the bird may adopt a variety of gliding strategies – to minimize sinking speed, for example, or to maximize forward travel. Comparisons of the varying gliding capacities of birds of different shapes and size have revealed that, because of their better aerodynamics, larger birds such as vultures and albatrosses do better than smaller birds; for sustained gliding they have also learned how to extract better advantage from the air in which they fly.

The details of how birds exploit movements of the air to *gain* height in gliding pose difficult problems of aerodynamics, but several markedly different methods have been distinguished. The upwards

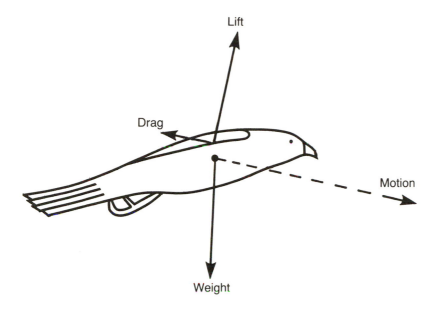

Fig. 6.13 *Gliding flight.*

component of wind passing over rising ground, such as a cliff, provides a relatively straightforward means, although the extent of the possible gain is clearly limited by the terrain. A less restricted method makes use of thermal up-risings of air associated with localized heating of the ground by the Sun. Large land-birds including eagles and vultures can soar thousands of feet by gliding upwards in a spiral path contained within the thermal current; by gliding downwards roughly in a straight line until the next thermal is encountered, Fig. 6.14, they can cover large distances with the minimum expenditure of energy. Interestingly, the time-table for the onset of thermal soaring during the course of a hot day is determined by aerodynamic capacity, and a definite order of launching has been observed extending from chcels having low wing-loadings to adjutants having wing-loadings almost three times as large.

Without the advantage of ground-based thermals, albatrosses have developed a technique that depends on the increase of speed with

Fig. 6.14 *Thermal soaring: the dotted lines indicate uprising currents of air and the full lines the path of the bird.*

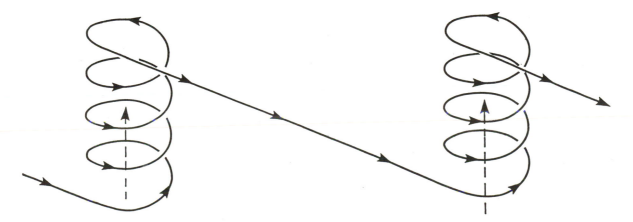

Fig. 6.15 *Gliding flight of an albatross over the sea: dotted lines indicate wind-speed, increasing with height.*

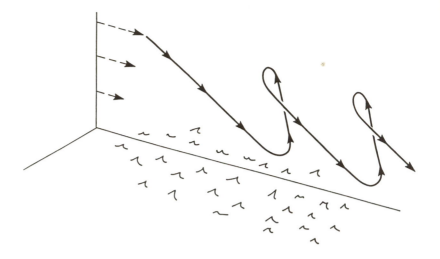

Fig. 6.15 *Gliding flight of an albatross over the sea: dotted lines indicate wind-speed, increasing with height.*

height of a horizontal wind, Fig. 6.15. In a downwind sinking glide, the albatross's speed over the sea is enhanced by the wind speed, but a time comes when there is no more height to lose. The albatross, then gliding at a relatively high airspeed, turns into the wind, which quickly increases lift and makes a climb possible. Although the flight path is now upwind, the increasing wind speed encountered as the albatross gains height not only yields a steeper climb but also ensures that the upwind phase of the manoeuvre covers less ground than the preceding downwind phase. When enough height has been gained, the albatross once again turns downwind and repeats the cycle by accelerating into a sinking glide. This cycle of events can continue for hours without recourse to flapping the wings.

In flapping flight, the ability of the bird to change the shape of its wings during the course of a beat assumes great importance. Wings can bend, twist and extend as well as beat, and groups of feathers can vary their relative positions. Such adaptability poses severe problems in aerodynamic theory, but makes possible modes of flying as different as the hovering flight of a humming bird and the fast forward flight of a swift. And as individual birds often execute a wide range of manoeuvres from take-off to landing, a correspondingly wide variation in the aerodynamic function must be expected.

Fig. 6.16 illustrates the changes in lift and drag forces acting on a section of a condor's wing during one beat of forward flight. The lift that keeps the bird airborne is generated during the powerful downstroke of the wing; the section offers a positive angle of attack relative to the dotted line marking the direction of motion, and the lift force, acting at a right angle to that line, provides a forward component for propulsion. On the up-beat, the wing section is rearranged to take up a small or zero angle of attack, so that the principal aerodynamic force is the considerably smaller drag. The net effect over a beat is to provide both a vertical force to maintain height and a forward thrust for propulsion. Fig. 6.17 shows an ingenious toy bird of French origin, driven by an elastic band, which managed to beat its wings with some success.

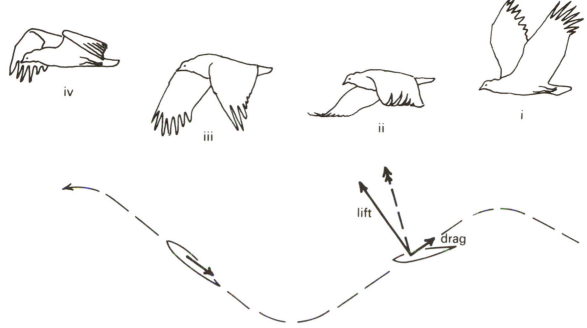

Birds that hover display even greater gymnastic virtuosity. Fig. 6.18 illustrates the action of a humming bird. The longitudinal axis of the body is not far from the vertical and the sections of the wings sweep out a figure-of-eight. During the forward stroke, a characteristic section adopts a positive angle of attack relative to the direction of its

Fig. 6.16 *The aerodynamics of a condor's wing in forward flight: the dotted line marks the path of a section of the wing (on an exaggerated scale). From McNeill Alexander, 1982.*

Fig. 6.17 *Flapping flight of a mechanical bird.*

Fig. 6.18 *Hovering flight of a humming bird. The full line marks the figure-of-eight path of a wing section. The black shapes show two positions of the section, one of them twisted into reverse during the back-stroke: both positions present a positive angle of attack and generate positive lift. From Hertel, 1966.*

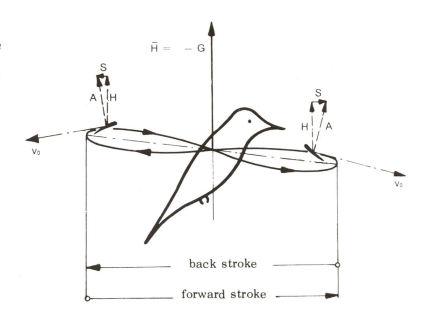

motion, thereby generating a continuous upwards component of lift. During the backwards stroke, the wing is twisted about its axis so that the same edge again leads the motion, and a positive angle of attack continues to provide an upwards component of lift. The horizontal component of the aerodynamic force actually reverses during the course of a complete cycle, thus forcing a forward and backward quiver of the whole bird, but the resulting amplitude is small.

Detailed theoretical and observational studies continue to reveal new aspects of the highly complicated aerodynamics involved in bird and insect flight. Although the distinctive actions of gliding, forward flight and hovering entail entirely different wing motions, intermediate patterns of flight are also possible. Yet other aspects of flight continue to be identified and explored by researchers: studies of control and stability, for example, and the effect of spread feathers present many lively areas of aerodynamic investigation. The structure of a feather, too, is a remarkable piece of engineering design; it consists of a thin-walled shaft of strong but light material, tapering in section from base to tip and containing a foam-like solid substance, an arrangement beautifully adapted to its purpose.

To complete the section, and for their own interest, Fig. 6.19 reproduces some of the flight patterns captured in a remarkable series of photographs.

Fig. 6.19 *Flight captured. (a) A white-bellied sea eagle at take-off; (b) a hovering cock nightjar showing reversal of the left wing. From Hoskin and Lane, 1970.*

Fig. 6.19 (c) *A Montagu's harrier*
alighting at its nest (a photograph
adopted by 193 Squadron of the
RAF as its crest); (d) a robin
showing a spectacular cascade of
feathers; (e) a sparrow-hawk
generating high lift, showing
feathers at 3 and 4 that cause
turbulence and improve the
boundary layer. (c) and (d) from
Hoskin and Lane, 1970; (e) from
Hertel, 1966.

c

d

e

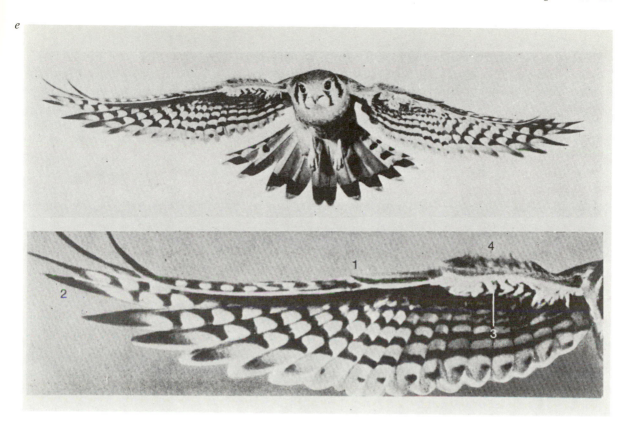

The human machine

The human skeleton is an elaborate assembly of links and joints. In engineering terms, it possesses more than 200 degrees of freedom, each providing a limited amount of relative motion between the connected parts. All are controlled by the nervous system. When one reflects that robots operate with up to about ten degrees of freedom, and that the task of coordinating so small a number requires extensive computer power, the full measure of human mobility becomes apparent.

Fig. 6.20(*a*) shows a skeleton made of steel, having joints that reproduce the relative motion of the human version. Close copies of this kind are used to study not only the kinematics of the skeleton but also the dynamics of motion in complicated events such as car crashes. Provided the distribution of mass is also a good copy of an actual body, much useful information can be deduced about the effect of seat-belts and other protective structures.

Coverings of the human body, although not normally regarded as the province of the engineer, sometimes involve considerable mechanical ingenuity. Fig. 6.20(*a*) also shows an exact replica of the spacesuit used by Colonel Buz Aldrin during the first landing on the Moon in 1969. Inevitably, the air pressure that had to be contained within the suit caused a general stiffening of the enclosing fabric, and so the design included specially flexible sections at the important joints to provide mobility. Fig. 6.20(*b*) shows an earlier design, free

Fig. 6.20 (a) *A steel model of a human skeleton and a replica of the space suit used by Colonel Buz Aldrin during the first Moon landing, and (b) a 16th century suit of armour protecting a 20th century occupant.*

a

b

from the complications of internal pressure, but with an equally pressing demand on the mobility of the 16th century occupant. The careful construction of the various gusseted joints, made up of overlapping pieces that could slide over each other, offered a combination of protection and relative movement that must have been a source of considerable comfort to its ancient owner: it also allowed the current occupant to raise his visor and walk away in full command of his progress.

Detailed studies of the way we move have revealed many interesting and often surprising aspects of the subject. Consider walking, which is a highly organized activity. As the left leg is raised at the hip, the left knee bends slightly, the left foot leaves the ground as the right foot supports the weight, the centre of mass of the body first rises then falls along an arc as it moves forwards, and the right arm swings forwards in phase with the left foot as the left arm swings back. When the left foot reaches the ground, the load is taken off the right foot, which remains briefly in contact with the ground before the next stride is taken. The cycle of events is then repeated.

At all times, one foot remains in contact with the ground as the weight of the body is transferred cyclically from one foot to the other. Measurements show that in a fast walk a man weighing 700 N (12 stone) presses on the ground with a maximum vertical force of 950 N during the loading of one foot, which is more than two and a half times the load corresponding to standing still; much higher multiplications occur during running or jumping. If one assumes that in a normal walking gait the arc followed by the centre of mass of the body is approximately circular, centred at the foot in contact with the ground, a simple calculation shows that the fastest forward speed that can be achieved by a normal adult man without his foot leaving the ground is between 5 and 6 miles per hour. To attain higher speeds, the force the ground would have to exert on the foot would have to pull *downwards* in order to provide the corresponding centripetal acceleration of the centre of mass. However, by varying this gait to flatten the natural rise and fall of the body, competitive walkers successfully avoid the limitations of the circular-arc theory.

Running is different. The loaded foot is used to propel the body forwards and upwards so that for part of the cycle the runner is entirely airborne. During this interval, the leading leg reaches forward so that the foot hits the ground to begin the next phase of the cycle from a bent-leg position. The forces involved at the foot and through the leg have to provide for much larger accelerations than those arising in walking, and the elasticity of the leg, the shoe and the ground assume a correspondingly greater importance. Various studies have sought to improve athletic performance by combining such factors in the best possible way – an interesting approach to a stable and controllable spring-heeled Jack!

Engineers have developed some remarkable instruments to measure the dynamical effects of human locomotion. The force platform shown in Fig. 6.21 is a sensitive weighing machine capable of measuring the forces caused by standing, walking, running or jumping

a

Fig. 6.21 (a) *A Kistler force-platform supporting a young volunteer; (b) variations in the vertical force on the platform caused by the pumping action of the heart.*

b

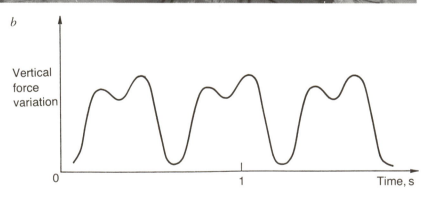

on it. By means of a clever arrangement of load cells, it can register not only the vertical component of the resultant force but also its two horizontal components, its point of application and the moment about a given vertical axis. And it can do so even if all the action is completed very quickly, as when a runner in full stride treads on the platform for a fraction of a second. Such information is valuable not only in sporting activities but also in orthopaedics, rehabilitation and studies of the nervous system.

In the photograph, Fig. 6.21(*a*), a young volunteer is shown standing on the platform. To stand *absolutely* still is impossible, and the inevitable slight swaying changes the load on the platform to a degree that can readily be detected; in clinical diagnosis of the

neuromuscular system, its subsequent analysis can be important. But
the demonstration revealed an even smaller effect. When the heart
pumps blood upwards through its exit or aorta valve, the upwards
momentum of the blood is accompanied by a tiny increase in pressure
on the platform, and this regular variation was also detectable on the
display screen. The accompanying drawing illustrates a characteristic
effect.

Fig. 6.22(a) shows another volunteer in transit over the platform.

a

b

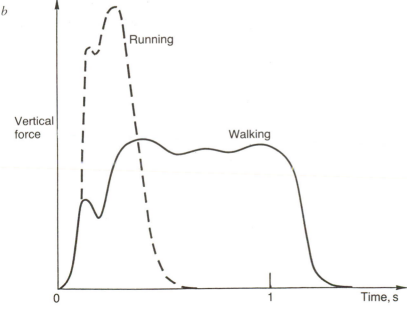

Fig. 6.22 (a) *Another volunteer in
transit over the platform;*
(b) *variations in the vertical force
when walking (full line) and
running (dotted line), at a lower
magnification than Fig. 6.21.*

With a reduced sensitivity of the instrument, a series of tests demonstrated the ways in which the vertical component of the foot-force varied as he changed his gait from walking to running. The drawings illustrate the differences. In walking, the foot remains in contact with the platform for up to about one second, and a distinct pattern of force variation, corresponding to heel-strike and toe-off, can be detected. In running, the interval of contact is much less, and a sharp peak is registered as the foot hits the ground. Detailed studies of such records have contributed valuable information about the forces transmitted through the body, both for clinical diagnosis and for the design of artificial joints.

Another development, shown in Fig. 6.23, displays the resultant force on the foot as a vector. In this ingenious arrangement, the signals generated within the force platform are received in the form of electrical voltages by a small computer, which returns the information

a

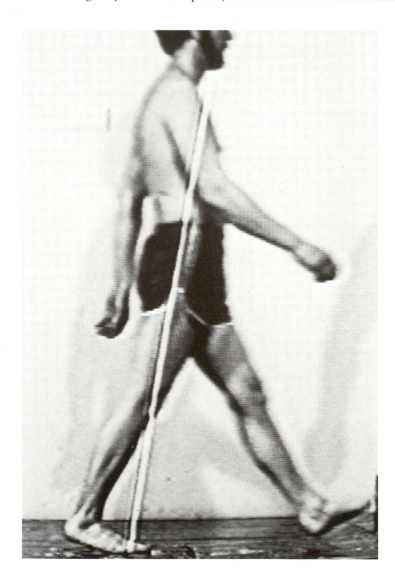

Fig. 6.23 *The photograph is taken from a television screen. The force-platform over which the subject is walking supplies information to a computer which enables the force to be represented as a vector on the screen. Note how the vector passes through the knee and hip joints.*

for display as a vector on a television screen, accurately superimposed on the picture of the moving leg. The composite result gives a clear view of the magnitude, the direction and the precise point of application of the force acting on the foot throughout the action. One then has to hand an immediate moving picture of the force at the foot, and valuable information can be deduced about the rather complicated loading on individual joints such as the knee and hip.

As well as investigating the *forces* involved in various forms of human locomotion, researchers and clinicians have made extensive studies of the *kinematics* of human movements. In view of the number of joints and limbs involved, all moving in three dimensions, it can be appreciated that this is not a simple task, and the traditional method of filming patients moving in front of a grid marked on a wall makes the analysis of the record a lengthy affair. The machine shown in Fig. 6.24 does things much more efficiently.

Behind the three windows in the box are three identical octagonal prisms whose sides are reflecting mirrors; they are driven around in accurate synchronization at 1800 rev/min, the outer two about vertical axes and the central one about a horizontal axis. To track a moving point, a small and specially designed glass prism is attached and, provided this marker moves within a generous field of illumination provided by the machine, all will be well. The rays of light reflected from the marker are detected photo-electrically when

Fig. 6.24 *Making a record of arm movements in Newcastle University's Biomedical Engineering Laboratory. (a) The displacements of the small glass prisms attached to the arm are tracked and recorded by means of reflected light and associated computery: the record (b), p.180, shows how they vary with time. Courtesy of Movement Techniques Ltd.*

a

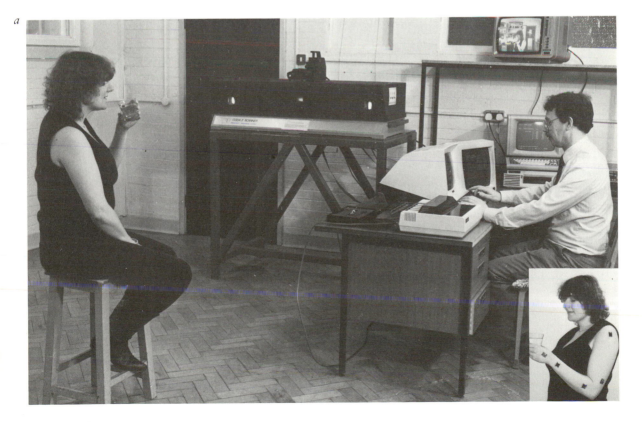

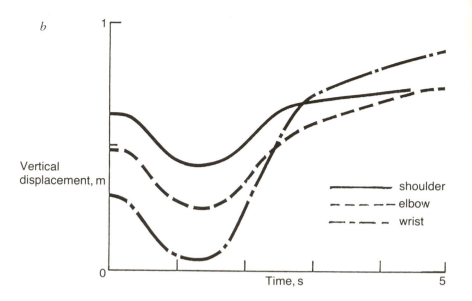

b

they impinge upon the rotating mirrored surfaces, and the precise positions of the receiving mirrors at that instant enable the three coordinates of the marker (x, y, z) to be calculated almost instantaneously by an associated and suitably programmed computer. The output, showing any two of the three coordinates or a single coordinate as a function of time, can be displayed on a television screen or drawn out for the record by a pen or a plotter. By this means, the movements of up to eight separate colour-coded markers attached to different sites on the subject's body can be tracked simultaneously – and since the results become available at once the investigator is spared much painstaking analysis.

Spare parts

With more and better information to hand about human forces and movements, the combined efforts of doctors and engineers on behalf of patients have achieved outstanding successes. The replacement of a defective hip joint by an artificial version, illustrated diagrammatically in Fig. 6.25(a), has become a familiar operation of great benefit to many who would otherwise have remained seriously disabled; the requirements of strength, lubrication, geometry, chemistry, reliability and life are severe, and the extent of the usage is a measure of its success. The artificial knee joint shown in Fig. 6.25(b) is a more complicated device that has to allow for the involved rolling and sliding kinematics of the natural knee, as well as for the transmission of force; as with the hip joint, its insertion can restore mobility that would otherwise be lost.

Another field in which biomedical engineering continues to advance is in the repair of damaged hearts. Fig. 6.26(a) illustrates the pumping action of the heart. As blood returns to the heart after circulation through the body, it is received by the right atrium and then passes through the tricuspid valve T to the right ventricle. From there it is

a

b

Fig. 6.25 (a) *An artificial hip joint showing the metal stem for fixing in the femur, and the acetabular cup with high-density polyethylene liner for fixing in the pelvic bone: the combination is a carefully designed ball joint.* (b) *An artificial knee joint with metal-to-polythene bearing surfaces designed to reproduce the kinematics of the natural joint and to optimize stresses.*

pumped through the pulmonary valve P to the lungs, where it is oxygenated before returning to the left atrium. Thereafter it flows through the mitral valve M into the left ventricale and is pumped through the aortic valve A into the aorta for the next round of circulation. In normal conditions the heart beats about 70 times a minute, and the flow rate is about 5 l/min; year in and year out, heart valves open and close about 40 million times to permit the flow of about 3 million litres of a complicated fluid. If a valve anywhere in the circuit becomes defective, the whole circulation is affected and it may become necessary to replace it by an artificial valve. To reproduce the superb performance of a healthy natural valve, working with a gentle folding action, poses difficult problems of engineering design, materials science and surgery. Fig. 6.26(*b*) shows some of the devices that have been used.

The difficulties of implanting an artificial replacement become much greater where the whole heart is concerned. Although

Fig. 6.26 (a) *The passage of blood through the heart: the compartments are atrium A and ventricle V; the valves are tricuspid T, pulmonary P, mitral M and aortic A. R is right and L is left. From Green, 1984.*
(b) *Some artificial heart valves.*

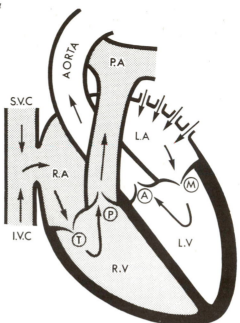

engineering pumps have been developed for a vast range of applications, the extraordinary requirements of the implanted artificial heart – size, weight, power supply as well as the full spectrum of biological compatibilities – have so far restricted application. Instead, greater success has been achieved through heart transplants from recently dead donors. But, outside the body, extra-corporeal machines, free from many of the limitations of artificial implants, play a vital rôle in clinical practice. The heart–lung machine, Fig. 6.27(*a*), enables the operating team to maintain the patient's circulation and breathing during operations on the heart; it includes pumps which drive the blood through oxygenators having heat-transfer capabilities, built-in monitors and fail-safe devices, all kept under strict supervision as the operation proceeds. The artificial kidney machine, Fig. 6.27(*b*), is used by patients when their natural kidneys can no longer remove waste products from the blood or regulate the chemical balances of body-fluids. Its central element is the

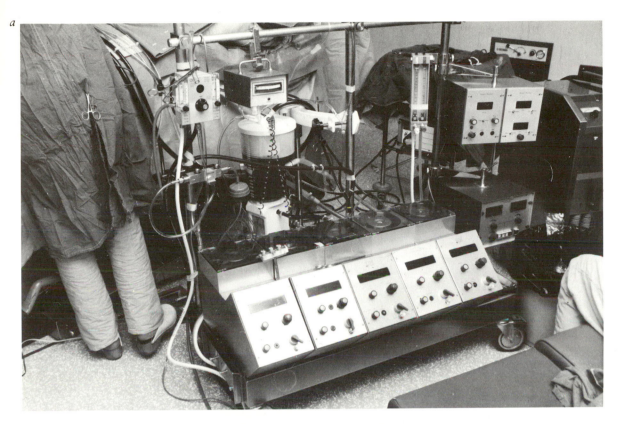

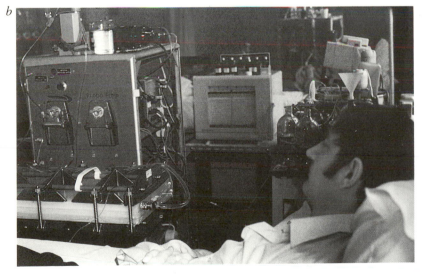

Fig. 6.27 (a) *A heart–lung machine in the operating theatre. Courtesy of Freeman Hospital, Newcastle upon Tyne.* (b) *An artificial kidney machine in use: the dialyser is at bottom left.*

dialyser in which blood from the body is passed over a membrane porous enough to allow the transport of impurities but impermeable to blood proteins and cells; a cleansing liquid or dialysate flowing on the other side of the membrane removes the unwanted elements. These machines, which continue to be the subject of much research and development, have made remarkable contributions to clinical practice.

In the field of artificial limbs for the disabled, there have been striking advances as the dynamics and control of the devices have improved. Extensive research and development have dealt not only with methods of reproducing natural kinematics but also with the ways in which power can be supplied and controlled. Fig. 6.28 shows an electrically powered artificial hand designed for use by young amputees. A socket made of light but strong plastic material, carefully adjusted to the individual for maximum comfort, is fitted to the stump of the arm below the elbow. Electrodes placed in contact with the skin detect the tiny voltages generated by the muscles when the user wishes to use the hand, and these signals are transmitted to an electronics package which switches on a miniature electric motor contained within the hand. By this means, the thumb and two opposing fingers can be opened and closed with a degree of control that makes even delicate gripping possible. Alternatively, a shoulder harness can be fitted which operates the switches mechanically and this, too, provides a high level of control.

a

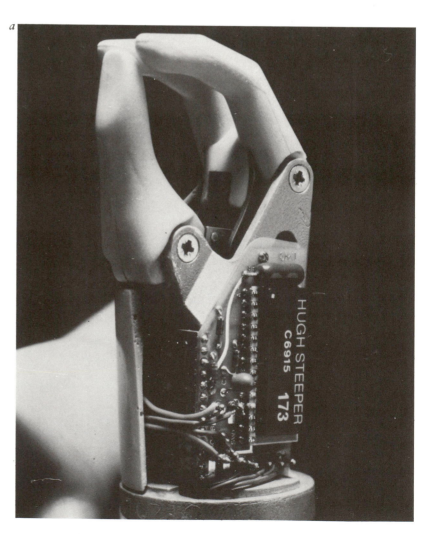

Fig. 6.28 *The Systemteknik/ Steeper powered hand (a) in detail, (b) in use. Courtesy of Hugh Steeper (Roehampton) Ltd.*

b

a

b

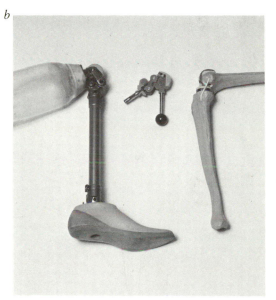

The artificial leg illustrated in Fig. 6.29 is a passive mechanical device for the use of above-the-knee amputees. To provide a successful knee-action for a range of actions is a difficult design task. The mechanism must be stable under load when standing and walking, capable of bending and self-centring during movement, and it must also reproduce as closely as possible the kinematics of the natural knee. The solution shown incorporates a four-bar linkage together with a restraining spring. When standing still, the user can remain perfectly stable without any tendency of the knee joint to bend; when walking, the loading on the joint will first release the required bending action, then once again support the user stably as he or she moves into the next stride.

Fig. 6.29 *Components of the Otto Bock artificial leg: (a) the partly sectioned knee joint, and (b) the assembly.*

Man and machines

Man's ability to make good use of the forces and materials available in nature is nowhere more apparent than in the immense range of machines at our disposal; the material conditions of life have been transformed by the skills of the inventor, designer and maker. But their story would not be complete without reference to the skills of the user as well. Although many machines operate obediently at the turn of a switch, many others call for a high degree of expertise if the results are to be as intended.

The humble roller-skate is a case in point, as anyone who has not acquired the necessary skill will readily testify; although Newton's laws still apply in full measure, it is not an easy matter for the unitiated to keep gravity at bay. Bicycling, too, calls for a good relationship with the principles of mechanics, even on the straight and level, while those who practise amazing feats on wheels seem to have arrived at a higher level of mechanics altogether.

Sometimes, the exercise of skill is not so obvious. The successful installation and commissioning of a large production machine, or a turbine generator set, requires the sustained effort of a team having high levels of experience and expertise. Careful procedures for assembly, inspection and test, sometimes for months on end, must be completed before the machine can be handed over in safe working order to its operators.

In contrast, the combined operations of man and machine are more widely recognized in the dramatic environments of space or the oceans. The manipulator arms, Fig. 6.30(a), that can be seen extending from space-vehicles are capable of performing elaorate tasks under the control of the on-board astronaut, whilst at the bottom of the sea unmanned submersibles, carrying their own eyes in the form of television cameras, can seek out and retrieve even small objects from the ocean floor under the directions of the mother-ship.

a

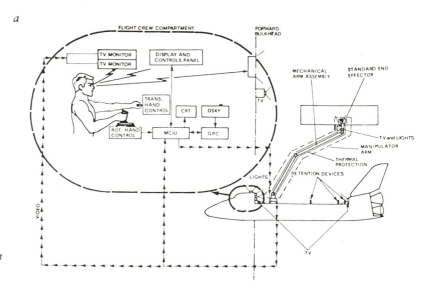

Fig. 6.30 *Remote-controlled manipulator arms at work (a) in space, (b) in the core of a nuclear reactor (courtesy of CEGB), (c) in the lecture theatre.*

b

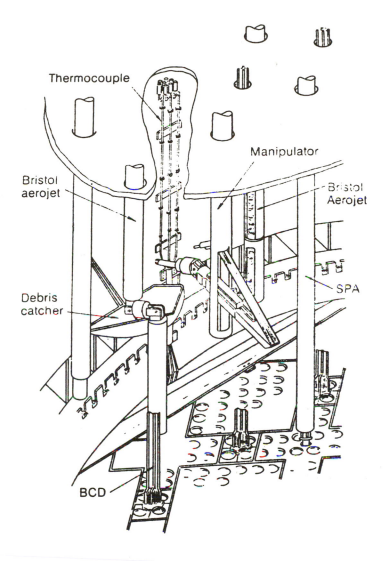

Thermocouple

Manipulator

Bristol
aerojet

Bristol
Aerojet

Debris
catcher

SPA

BCD

c

The development of humanly controlled manipulators for use in hazardous environments is important also at sites subject to high levels of nuclear radiation, such as the cores of nuclear reactors in power stations, Fig. 6.30(*b*). Inspection of these uninhabitable chambers, together with tasks of maintenance and repair, is an essential part of their operation, and skilled teams equipped with television cameras and manipulator arms have cleverly developed the means of carrying out these functions. Fig. 6.30(*c*) shows a small manipulator designed to handle radioactive specimens in a research laboratory. The human controller has the advantage of direct vision but must be able to position the slave arm with great dexterity. The mechanics of his hand movements determine the electrical signals issued to the slave, and, in view of the three-dimensional nature of the movements, a high level of operational skill is required, even for the less technical operation shown in the photograph.

The most familiar man–machine combination is the road vehicle. The driver usually has five control devices at his disposal – steering wheel, clutch, gear-change, accelerator and brake. The clutch and gear-change provide the driver with the means of adjusting the engine speed in relation to the road speed. Together with the accelerator, which controls the rate of fuel delivered to the engine, a wide range of road speed and acceleration is available to meet variable driving conditions. Whether the driver can remain within the working limits of the engine depends, naturally, on his or her skill. Fortunately, a legally acceptable level of performance is within the reach of most potential drivers. But to obtain the best possible result from a particular machine, as determined for example by fuel economy on a given run or by minimum time on a racing track, presents a considerably more difficult challenge.

Fig. 6.31 *The RAF Red Arrows, 1985.*

a

Fig. 6.32 *Red Arrows in flying formations.*

b

When the motion takes place along complicated three-dimensional curves at speeds in excess of 600 miles per hour, a few feet away from other accompanying vehicles, the results are bound to be spectacular. Following an RAF tradition of formation aerobatics, the famous Red

Arrows were formed in 1965, and their pilots have been demonstrating their remarkable flying skills ever since. The current (1985) team is shown in Fig. 6.31, and some pictures of their formations in Fig. 6.32. The aircraft used is the British Aerospace Hawk, having a wing-span of 31 ft and a maximum speed of 633 miles per hour in level flight, and its engine is a mark of the Rolls-Royce Ardour jet engine shown in Fig. 5.28. During the lecture, Flight Lieutenant Simon Bedford, on the extreme left of the group photograph, provided a running commentary on a film of some of the Red Arrows aerobatic exploits. He also explained how it was all done, though few of the audience could believe that such a memorable exhibition of Newton's laws was quite as simple as an expert made it sound.

Conclusion

Conclusions should always be brief. The Royal Institution's Christmas Lectures are intended to capture the imagination of young people, but there is no doubt that they also capture the imagination of all who take part, behind as well as in front of the scenes. It was tremendous fun.

Further reading

CHAPTER 1
A.F. Burstall. *A History of Mechanical Engineering*. Faber and Faber, 1963
R.J. Hulsizer and D. Lazarus. *The World of Physics*. Addison Wesley, 1977
S. Lilley. *Man, Machines and History*. Laurence and Wishaw, 1965

CHAPTER 2
R.N. Arnold and L. Maunder. *Gyrodynamics*. Academic Press, 1961
J.P. den Hartog. *Mechanics*. Dover, 1961
J.L. Meriam. *Dynamics*. John Wiley, 1980
J.L. Synge and B.A. Griffith. *Principles of Mechanics*. McGraw-Hill, 1959

CHAPTER 3
R.E.D. Bishop. *Vibration*. Cambridge University Press, 1979
J.P. den Hartog. *Mechanical Vibrations*. McGraw-Hill, 1947
G.B. Warburton. *The Dynamical Behaviour of Structures*. Pergamon, 1964

CHAPTER 4
J. Aleksander and P. Burnett. *Reinventing Man*. Pelican, 1984
S. Bennett. *A History of Control Engineering*. Peter Peregrinus, 1979.
G.J. Thaler and R.G. Brown. *Analysis and Design of Feedback Control Systems*. McGraw-Hill, 1960
J. Vertut and P. Coiffet. *Teleoperation and Robotics*. Kogan Page, 1985.

CHAPTER 5
C.B. Daish. *The Physics of Ball Games*. English University Press, 1972
C.H. Gibbs-Smith. *Flight Through the Ages*. Crowell, 1974
R. Hawkey. *Sport Science*. Hodder and Stoughton, 1981
B.S. Massey. *Mechanics of Fluids*. Van Nostrand Reinhold (UK), 1979
A.H. Shapiro. *Shape and Flow*. Heinemann, 1974

CHAPTER 6
J. Gray. *Animal Locomotion*. Weidenfeld & Nicholson, 1968.
J.H. Green. *Basic Clinical Physiology*. Oxford University Press, 1984
H. Hertel. *Structure – Form – Movement*. Reinhold, 1966
M. Hildebrand, 1959. *J. Mammology* 40, 481–95
E. Hoskin and F.W. Lane. *An Eye for a Bird*. Hutchinson, 1970
L.M. Klauber, 1944. *Trans. San Diego Soc. Nat. Hist.* 10, 91–126
R. McNeill Alexander. *Locomotion of Animals*. Blackie, 1982
R. McNeill Alexander and G. Goldspink. *Mechanics and Energetics of Animal Locomotion*, p.239. Chapman Hall, 1977
W. Nachtigall, 1960. *Z vergl. Physiol.* 43, 48–118

Index